Type-2 Fuzzy Neural Networks
and Their Applications

Rafik Aziz Aliev • Babek Ghalib Guirimov

Type-2 Fuzzy Neural Networks and Their Applications

 Springer

Rafik Aziz Aliev
Azerbaijan State Oil Academy
Baku, Azerbaijan

Babek Ghalib Guirimov
SOCAR
Baku, Azerbaijan

ISBN 978-3-319-38159-6   ISBN 978-3-319-09072-6 (eBook)
DOI 10.1007/978-3-319-09072-6
Springer Cham Heidelberg New York Dordrecht London

Printed on acid-free paper

Springer is part of Springer Science+Business Media (www.springer.com)

Dedicated to the memory of my wife Aida Alieva
Rafik Aliev
To my father
Babek Guirimov

Preface

Theory and practice of artificial intelligence (AI) have been widely used in many various areas of human activity. When there was intense competition between various methodologies linked to AI, it was Prof. L. Zadeh's perception at the time that more could be gained by cooperation than by claims and counter-claims of superiority. A strong need has appeared for a new approach, theory and technology for the development of intelligent systems with high level of machine intelligence quotient (MIQ).

To answer this question Prof. L. Zadeh suggested so-called soft computing (SC) technology, representing a consortium of such intelligent paradigms as fuzzy logic (FL), neurocomputing (NC), evolutionary computing (EC), probabilistic computing (PC), and chaotic computing (CC) that have allowed the solution of many important real-world problems, which traditional methods of artificial intelligence were unable to do. The main feature of these methodologies in common is that they are more tolerant to imprecision, uncertainty, and partial truth than their hard-computing-based counterparts.

An essential characteristic of soft computing is that its constituent methodologies are, for the most part, complementary rather than competitive. Thus, the basic tenet of soft computing is that, in general, better results can be achieved by employing the constituent methodologies in combination than in a stand-alone mode. Today, a combination that has the highest visibility is that of so-called neuro-fuzzy systems. Another combination which is growing in visibility is neuro-fuzzy-genetic systems.

Neuro-fuzzy hybridization results in a hybrid intelligent system that synergizes these two techniques by combining the human-like reasoning style of fuzzy systems with the learning and connectionist structure of neural networks. The lack of interpretability of neural networks on the one hand and the poor learning capability of fuzzy systems on the other hand are similar problems that limit the application of these tools. Neuro-fuzzy systems are hybrid systems which try to solve this problem by combining the learning capability of connectionist models and the interpretability property of fuzzy systems. In case of a dynamic work environment, the automatic knowledge base correction in fuzzy systems becomes necessary. On the other hand, artificial neural networks are successfully used in problems connected to knowledge acquisition using learning by examples with the required degree of precision. Incorporating neural networks in fuzzy systems for fuzzification, construction of fuzzy rules, optimization and adaptation of fuzzy knowledge base, implementation of fuzzy reasoning, and defuzzification is the essence of the neuro-fuzzy approach. As a result, neural networks become more transparent, while fuzzy systems become capable of learning. Expert knowledge can be incorporated into the structure of the neuro-fuzzy system. At the same time, the connectionist structure avoids fuzzy inference, which entails a substantial computational burden.

Ordinary neuro-fuzzy systems (i.e. type-1 neuro-fuzzy systems) have been successfully used in a wide range of applications. To design ordinary neuro-fuzzy systems, knowledge of human experts and experimental data are needed for construction of fuzzy rules and membership functions based on available linguistic or numeric information. However, in many cases the available information or data are associated with various types of uncertainty which should be taken into account. This uncertainty can be captured well by using higher order fuzzy sets. Hence, an effective way is to employ type-2 fuzzy sets, which augment fuzzy models with expressive power to develop models that efficiently capture the factor of uncertainty. In this regards, fuzzy type-2 neuro-fuzzy systems can represent and handle uncertain information more effectively than fuzzy type-1 neuro-fuzzy systems and contribute to the robustness and stability of the inference. Type-2 fuzzy sets having offered additional degrees of freedom in combination with neural networks being viewed as parallel computational models with adaptive nature make it possible to directly and more effectively account for model's uncertainties produced by different sources. The concept of type-2 fuzzy sets was initially created by Prof. L. Zadeh as an extension of ordinary fuzzy sets. Then Mendel and Karnik have developed a theory of type-2 fuzzy systems.

In spite of the intensive development of theory and design methods of type-2 neuro-fuzzy systems, the concept of type-2 neuro-fuzzy system is still in its initial stages of crystallization. There is little progress in the area of type-2 fuzzy rule extraction, merging type-2 fuzzy logic system with other constituents of soft computing, namely, with evolutionary computing, etc.

In this view, this book deals with the theory, design principles, and application of hybrid intelligent systems using type-2 fuzzy sets in combination with other paradigms of soft computing technology such as neuro-computing and evolutionary computing.

This book is organized into five chapters. The first chapters is a succinct exposition of the basics of fuzzy logic, with emphasis on those parts of fuzzy logic which are of prime relevance to neuro-fuzzy approach. Here foundations of fuzzy sets theory, including basic concepts and properties of type-1 and type-2 fuzzy sets, elements of fuzzy mathematics, and fuzzy relations are given. This chapter also contains elements of fuzzy logic (in narrow sense of L. Zadeh), which is viewed as a generalization of the various multi-valued logics. This chapter includes compositional rules of inference, choice of fuzzy implications, and formalization of fuzzy conditional inference using the fuzzy implications suggested by the authors.

Chapter 2 deals with comparative analysis of methods of evolutionary computing and its merging with type-2 fuzzy neural networks. The models of neural networks and, especially, fuzzy and high-order fuzzy networks can be described by complex nonlinear, non-convex, and non-differentiable functions. As evolutionary optimization methods do not require any restrictive properties for the functions or the computational models to work with, they can be effectively used for training of parameters of fuzzy type-2 neural networks. The chapter includes main characteristics of genetic algorithms, particle swarm optimization (PSO), and differential evolution (DE) methods. An emphasis is put on the DE method and its application to the training of all types of neural networks.

Type-1 and type-2 fuzzy neural networks are subject of Chap. 3. This chapter includes for neuro-fuzzy computing, basic architectures and operations of fuzzy feed-forward and recurrent neural networks, several types of fuzzy logical neuron models, logic-oriented neural networks, general and interval type-2 fuzzy neural network's features and their training algorithms.

Chapter 4 contains elements of general and interval type-2 FCM clustering methods, interval type-2 fuzzy clustering using DE, and design of type-2 neural networks with clustering.

Chapter 5 describes application of type-2 neural networks to real-word problems in decision making, forecasting, control, identification and other areas.

The suggested book is devoted in its entirety to a systematic account of major concepts and methodologies of type-2 fuzzy neural networks and presents a unified framework that makes the subject more accessible to students and practitioners.

We would like to express our thanks to Prof. Lotfi A. Zadeh for his constant support, invaluable ideas, and advices.

Baku, Azerbaijan   Rafik Aziz Aliev
                                                                        Babek Ghalib Guirimov

Contents

Chapter 1
Fuzzy Sets

1.1 Type-1 Fuzzy Sets

Definition 1.1 *Type-1 fuzzy sets.* Let X be a classical set of objects, called the universe, whose generic elements are denoted x. Membership in a classical subset A of X is often viewed as a characteristic function μ_A from A to $\{0, 1\}$ such that

$$\mu_A(x) = \begin{cases} 1, & \textit{iff } x \in A \\ 0, & \textit{iff } x \notin A \end{cases}$$

where $\{0, 1\}$ is called a valuation set; 1 indicates membership while 0 – non-membership.

If the valuation set is allowed to be in the real interval [0,1], then A is called a fuzzy set (Aliev and Aliev 2001; Klir and Yuan 1995; Klir et al. 1997; Zadeh 1965; Zimmermann 1996), $\mu_A(x)$ is the grade of membership of x in A.

$$\mu_A : X \rightarrow [0, 1]$$

As closer the value of $\mu_A(x)$ is to 1, so much x belongs to A. A is completely characterized by the set of pairs:

$$A = \{(x, \mu_A(x)), \quad x \in X\}$$

Example The representation of temperature within a range $[T_1, T_2]$ by type-1 fuzzy and crisp sets is shown in Fig. 1.1a, b, respectively. In the first case we use membership function $[T_1, T_2] \rightarrow [0, 1]$ for describing linguistic concepts *"cold"*, *"normal"*, *"warm"*. In the second case right-open intervals are used for describing of traditional variable by crisp sets.

Fuzzy sets type-1 with crisply defined membership functions are called ordinary fuzzy sets.

© Springer International Publishing Switzerland 2014

R.A. Aliev, B.G. Guirimov, *Type-2 Fuzzy Neural Networks and Their Applications*, DOI 10.1007/978-3-319-09072-6_1

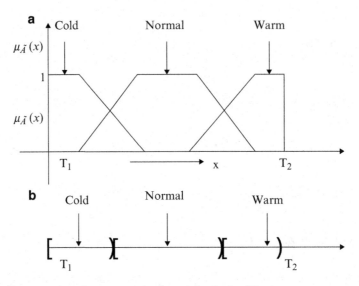

Fig. 1.1 Representation of temperature by fuzzy (**a**) and crisp (**b**) sets

Definition 1.2 *Equality of type-1 fuzzy sets.* Two fuzzy sets A and B are said to be equal if and only if

$$\forall x \in X \quad \mu_A(x) = \mu_B(x) \quad A = B.$$

Definition 1.3 *Support and crossover point of a type-1 fuzzy set. Singleton.* The support of a fuzzy set A is the ordinary subset of X that has nonzero membership in A:

$$\text{supp}A = A^{+0} = \{x \in X, \mu_A(x) > 0\}$$

The elements of x such as $\mu_A(x) = 1/2$ are the crossover points of A.

A type-1 fuzzy set that has only one point in X with $\mu_A = 1$ as its support is called a singleton.

Definition 1.4 *The height of a type-1 fuzzy set. Normal and subnormal sets.* The height of A is

$$hgt(A) = \sup_{x \in X} \mu_A(X)$$

i.e. the least upper bound of $\mu_A(x)$.

A is said to be normalized iff $\exists x \in X, \mu_A(x) = 1$. This definition implies hgt $(A) = 1$. Otherwise $\tilde{A}$ is called subnormal type-1 fuzzy set.

The empty set $\varnothing$ is defined as

$$x \in X, \mu_\varnothing(x) = 0, \text{ of course} \forall x \in X \mu_X(x) = 1$$

Definition 1.5 α-level type-1 fuzzy sets. One of important way of representation of fuzzy sets is α-cut method. Such type of representation allows us to use properties of crisp sets and operations on crisp sets in fuzzy set theory.

The (crisp) set of elements that belongs to the fuzzy set A at least to the degree α is called the α-level set:

$$A^\alpha = \{x \in X, \mu_A(x) \geq \alpha\}$$

$A^\alpha = \{x \in X, \mu_A(x) > \alpha\}$ is called a strong α-level set or a strong α-cut.

Now we introduce fuzzy set A_α, defined as

$$A_\alpha(x) = \alpha A^\alpha(x) \tag{1.1}$$

Then the original fuzzy set A may be defined as $A = \bigcup_{\alpha \in [0,1]} A_\alpha$. $\cup$ denotes the standard fuzzy union.

Definition 1.6 Convexity of type-1 fuzzy sets. A fuzzy set A is convex iff

$$\mu_A(\lambda x_1 + (1 - \lambda)x_2) > \min(\mu_A(x_1), \mu_A(x_2)) \tag{1.2}$$

for all $x_1, x_2 \in R, \lambda \in [0, 1]$, min denotes the minimum operator.

Alternatively, a fuzzy set A on R is convex iff all its α-level sets are convex in the classical sense.

Definition 1.7 The cardinality of a fuzzy set. When X is a finite set, the scalar cardinality $|A|$ of a fuzzy set A on X is defined as

$$|A| = \sum_{x \in A} \mu_A(x)$$

Sometimes $|A|$ is called the power of $|A|$. $\|A\| = |A|/|X|$ is the relative cardinality. When X is infinite, $|A|$ is defined as

$$|A| = \int_X \mu_A(x)\, dx$$

Definition 1.8 Fuzzy set inclusion. Given type-1 fuzzy sets $A, B \in F(X)$ A is said to be included in B ($A \subseteq B$) or A is a subset of B if $\forall x \in X, \mu_A(x) \leq \mu_B(x)$

When the inequality is strict, the inclusion is said to be strict and is denoted as $A < B$.

1.1.1 Representation and Construction of Fuzzy Sets

It was mentioned above that each fuzzy set is uniquely defined by a membership function. In the literature one can find different ways in which membership functions are represented.

List representation. If universal set $X = \{x_1, x_2, \ldots, x_n\}$ is a finite set, membership function of a type fuzzy set A on X $\mu_A(x)$ can be represented as table. Such table lists all elements in the universe X and the corresponding membership grades as shown below

$$A = \mu_A(x_1)/x_1 + \ldots + \mu_A(x_n)/x_n = \sum_{i=1}^{n} \mu_A(x_i)/x_i$$

Here symbol / (slash) does not denote division, it is used for correspondence between an element in the universe X (after slash) and its membership grade in the fuzzy set A (before slash). The symbol + connects the elements (does not denote summation).

If X is a in finite set then

$$A = \int_X \mu_A(x)/x.$$

Here symbol $\int_X$ is used for denoting a union of elements of set X.

Graphical representation. Graphical description of a fuzzy set A on the universe X is suitable in case when X is one or two-dimensional Euclidean space. Simple typical shapes of membership functions are usually used in type-1 fuzzy set theory and practice (Table 1.1).

Fuzzy n cube representation. All type-1 fuzzy sets on universe X with n elements can be represented by points in the n-dimensional unit cube – n-cube. Assume that universe X contains n elements $X = \{x_1, x_2, \ldots, x_n\}$. Each element x_1, $i = \overline{1, n}$ can be viewed as a coordinate in the n dimensional Euclidean space. A subset of this space for which values of each coordinate are restricted in $[0,1]$ is called an n-cube. Vertices of the cube, i.e. bit list $(0, 1, \ldots, 0)$ represent crisp sets. The points inside the cube define fuzzy subsets.

Analytical representation. In case if universe X is infinite, it is not effective to use the above considered methods for representation of membership functions of a type-1 fuzzy sets. In this case it is suitable to represent type-1 fuzzy sets in an analytical form, which describes the shape of membership functions.

There are some typical formulas describing frequently used membership functions in type-1 fuzzy set theory and practice.

Table 1.1 Typical membership functions

Type of membership function	Graphical representation	Analytical representation
Triangular MF		$$\mu_A(x) = \begin{cases} \dfrac{x-a_1}{a_2-a_1}\,r, & \text{if } a_1 \le x \le a_2, \\[2mm] \dfrac{a_3-x}{a_3-a_2}\,r, & \text{if } a_2 \le x \le a_3, \\[2mm] 0, & \text{otherwise} \end{cases}$$
Trapezoidal MF		$$\mu_A(x) = \begin{cases} \dfrac{x-a_1}{a_2-a_1}\,r, & \text{if } a_1 \le x \le a_2, \\[2mm] r, & \text{if } a_2 \le x \le a_3, \\[2mm] \dfrac{a_4-x}{a_4-a_3}\,r, & \text{if } a_3 \le x \le a_4, \\[2mm] 0, & \text{otherwise} \end{cases}$$
S-spaped MF		$$\mu_A(x) = \begin{cases} 0, & \text{if } x \le a_1, \\[2mm] 2\left(\dfrac{x-a_1}{a_3-a_1}\right)^2, & \text{if } a_1 < x < a_2, \\[2mm] 1-2\left(\dfrac{x-a_1}{a_3-a_1}\right)^2, & \text{if } a_2 \le x < a_3, \\[2mm] 1, & \text{if } a_3 \le x \end{cases}$$

(continued)

Table 1.1 (continued)

Type of membership function	Graphical representation	Analytical representation
Bell-shaped MF		$\mu_A(x) = c \cdot \exp\left(-\frac{(x-a)^2}{b}\right)$

For example, bell-shaped membership functions often are used for representation of type-1 fuzzy sets. These functions are described by the formula:

$$\mu_A(x) = c\exp\left(-\frac{(x-a)^2}{b}\right)$$

which is defined by three parameters, a, b and c.

In general it is effective to represent the important typical membership functions by a parameterized family of functions. The following are formulas for describing the six families of membership functions

$$\mu_A(x, c_1) = \left[1 + c_1(x-a)^2\right]^{-1} \tag{1.3}$$

$$\mu_A(x, c_2) = \left[1 + c_2|x-a|\right]^{-1} \tag{1.4}$$

$$\mu_A(x, c_3, d) = \left[1 + c_3|x-a|^d\right]^{-1} \tag{1.5}$$

$$\mu_A(x, c_4, d) = \exp\left[-c_4|x-a|^d\right] \tag{1.6}$$

$$\mu_A(x, c_5) = \max\{0, [1 - c_5|x-a|]\} \tag{1.7}$$

$$\mu_A(x, c_6) = c_6\exp\left[-\frac{(x-a)^2}{b}\right] \tag{1.8}$$

Here $c_i > 0$, $i = \overline{1,6}$, $d > 1$ are parameters, the value of a corresponds the elements of the relevant fuzzy sets with the membership grade equal to unity.

The graphical and analytical representations of frequently used membership functions (MF) are shown in Table 1.1.

The problem of constructing membership functions is problem of knowledge engineering.

There are many methods for estimation of membership functions. They can be classified as follows:

1. Membership functions based on heuristics.
2. Membership functions based on reliability concepts with respect to the particular problem.
3. Membership functions based on more theoretical demand.
4. Membership functions as a model for human concepts.
5. Neural networks based construction of membership functions

The estimation methods of membership functions based on more theoretical demand use axioms, probability density functions and so on.

1.1.2   Operations on Type-1 Fuzzy Sets

There exist three standard fuzzy operations: fuzzy intersection, union and complement which are generalization of the corresponding classical set operations.

Let's A and B be two fuzzy sets in X with the membership functions μ_A and μ_B respectively. Then the operations of intersection, union and complement are defined as given below.

Definition 1.9 *Fuzzy standard intersection and union.* The intersection ($\cap$) and union ($\cup$) of fuzzy sets A and B can be calculated by the following formulas:

$$\forall x \in X \quad \mu_{A \cap B}(x) = \min\left(\mu_A(x), \mu_B(x)\right)$$
$$\forall x \in X \quad \mu_{A \cup B}(x) = \max\left(\mu_A(x), \mu_B(x)\right),$$

where $\mu_{A \cap B}(x)$ and $\mu_{A \cup B}(x)$ are the membership functions of $A \cap B$ and $A \cup B$, respectively.

Definition 1.10 *Standard fuzzy complement.* The complement A^c of A is defined by the membership function:

$$\forall x \in X \quad \mu_{A^c}(x) = 1 - \mu_A(x).$$

As already mentioned $\mu_A(x)$ is interpreted as the degree to which x belongs to A. Then by the definition $\mu_{A^c}(x)$ can be interpreted as the degree to which x does not belong to $\tilde{A}$.

The standard fuzzy operations do not satisfy the law of excluded middle $A \cup A^c = X$ and the law of contradiction $A \cap A^c = \varnothing$ of classical set theory. But commutativity, associativity, idempotency, distributivity, and De Morgan laws are held for standard fuzzy operations.

For fuzzy union, intersection and complement operations there exist a broad class of functions. Function that qualify as fuzzy intersections and fuzzy unions are defined as T-norms and T-conorms.

Definition 1.11 *T-norms.* T-norm is a binary operation in [0,1], i.e. a binary function T from [0,1] into [0,1] that satisfies the following axioms:

$$T(\mu_A(x), 1) = \mu_A(x) \tag{1.9}$$

if $\mu_A(x) \leq \mu_C(x)$ and $\mu_B(x) \leq \mu_D(x)$ then

$$T(\mu_A(x), \mu_B(x)) \leq T(\mu_C(x), \mu_D(x)) \tag{1.10}$$
$$T(\mu_A(x), \mu_B(x)) = T(\mu_B(x), \mu_A(x)) \tag{1.11}$$
$$T(\mu_A(x), T(\mu_B(x), \mu_C(x))) = T(T(\mu_A(x), \mu_B(x), \mu_C(x))) \tag{1.12}$$

Here Eq. 1.9 is boundary condition, (1.10), (1.11), and (1.12) are conditions of monotonicity, commutativity and associativity, respectively.

The function T takes as its arguments the pair consisting of the element membership grades in set A and in set B, and yields membership grades of the element in the $A \cap B$

$$(A \cap B)(x) = T[A(x), B(x)] \quad \forall x \in X.$$

The following are frequently used T-norm-based type-1 fuzzy intersection operations:

Standard intersection

$$T_0(\mu_A(x), \mu_B(x)) = \min\{\mu_A(x), \mu_B(x)\} \tag{1.13}$$

Algebraic product

$$T_1(\mu_A(x), \mu_B(x)) = \mu_A(x) \cdot \mu_B(x) \tag{1.14}$$

Bounded difference

$$T_2(\mu_A(x), \mu_B(x)) = \mu_{A \cap B}(x) = \max(0, \mu_A(x) + \mu_B(x) - 1) \tag{1.15}$$

Drastic intersection

$$T_3(\mu_A(x), \mu_B(x)) = \begin{cases} \min\{\mu_A(x), \mu_B(x)\} & \text{if } \mu_A(x) = 1 \\ & \text{or } \mu_B(x) = 1 \\ 0 & \text{otherwise} \end{cases} \tag{1.16}$$

For four fuzzy intersections the following is true

$$T_3(\mu_A(x), \mu_B(x)) \le T_2(\mu_A(x), \mu_B(x)) \le T_1(\mu_A(x), \mu_B(x)) \\ \le T_0(\mu_A(x), \mu_B(x)) \tag{1.17}$$

Definition 1.12 *T-conorms.* T-conorm is a binary operation in $[0, 1]$, i.e. a binary function $S : [0, 1] \times [0, 1] \to [0, 1]$ that satisfies the following axioms:

$$S(\mu_A(x), 0) = \mu_A(x) \quad \text{(boundary condition)} \tag{1.18}$$

if $\mu_A(x) \le \mu_C(x)$ and $\mu_B(x) \le \mu_D(x)$ then

$$S(\mu_A(x), \mu_B(x)) \le S(\mu_C(x), \mu_D(x)) \quad \text{(monotonicity)} \tag{1.19}$$

$$S(\mu_A(x), \mu_B(x)) = S(\mu_B(x), \mu_A(x)) \quad \text{(commutativity)} \tag{1.20}$$

$$S(\mu_A(x), S(\mu_B(x), \mu_C(x))) = S(S(\mu_A(x), \mu_B(x)), \mu_C(x)) \quad \text{(associativity).} \tag{1.21}$$

The function S yields membership grade of the element in the set $A \cup B$ on the argument which is pair consisting of the same elements membership grades in sets A and B

$$(A \cup B)(X) = S[A(x), B(x)] \tag{1.22}$$

The following are frequently used T-conorm based fuzzy union operations.

Standard union

$$S_0(\mu_A(x), \mu_B(x)) = \max\{\mu_A(x), \mu_B(x)\} \tag{1.23}$$

Algebraic sum

$$S_1(\mu_A(x), \mu_B(x)) = \mu_A(x) + \mu_B(x) - \mu_A(x) \cdot \mu_B(x) \tag{1.24}$$

Drastic union

$$S_3(\mu_A(x), \mu_B(x)) = \begin{cases} \max\{\mu_A(x), \mu_B(x)\} & \text{if } \mu_A(x) = 0 \\ & \text{or } \mu_B(x) = 0 \\ 1 & \text{otherwise} \end{cases} \tag{1.25}$$

For four fuzzy union operations the following is true

$$S_0(\mu_A(x), \mu_B(x)) \le S_1(\mu_A(x), \mu_B(x)) \le S_2(\mu_A(x), \mu_B(x)) \\ \le S_3(\mu_A(x), \mu_B(x)) \tag{1.26}$$

Definition 1.13 Cartesian product of type-1 fuzzy sets. The Cartesian product of fuzzy sets $A_1, A_2, \ldots, A_n$ on universes $X_1, X_2, \ldots, X_n$ respectively is a fuzzy set in the product space $X_1 \times X_2 \times \ldots \times X_n$ with the membership function

$$\mu_{A_1 \times A_2 \times \ldots \times A_n}(x) = \min\{\mu_{A_i}(x_i) | x = (x_1, x_2 \ldots, x_n), x_i \in X_i\}$$

Definition 1.14 Power of fuzzy sets. m-th power of a type-1 fuzzy set A^m is defined as

$$\mu_{A^m}(x) = [\mu_A(x)]^m , \quad \forall x \in X, \ \forall m \in R^+ \tag{1.27}$$

where R^+ is positively defined set of real numbers.

Definition 1.15 Concentration and dilation of type-1 fuzzy sets.
Let A be fuzzy set on the universe:

$$A = \{x, \mu_A(x)/x \in X\}$$

Then the operator $Con_m A = \{(x, [\mu_A(x)]^m)/x \in X\}$ is called a *concentration* of A and the operator $Dil_n A = \{(x, \sqrt{\mu_A(x)}) /x \in X\}$ is called a *dilation* of A.

Definition 1.16 *Difference of type-1 fuzzy sets.* Difference of fuzzy sets is defined by the formula:

$$\forall x \in X, \ \mu_{A|-|B}(x) = \max(0, \mu_A(x) - \mu_B(x)) \tag{1.28}$$

$A| - |B$ is the fuzzy set of elements that belong to A more than to B.

Symmetrical difference of fuzzy sets A and B is the fuzzy set $A \nabla B$ of elements that belong more to A than to B:

$$\forall x \in X \ \ \mu_{A \nabla B}(x) = |\mu_A(x) - \mu_B(x)| \tag{1.29}$$

Let us mention that fuzziness is essentially different from randomness. In order to understand this, let us consider the following concept. Assume that $P(X)$ is a power fuzzy set of the universe X. Then the mapping $\prod : P(X) \to [0, 1]$ with the following properties:

$$\prod(\varnothing) = 0, \ \prod(X) = 1;$$
$$A \subseteq B \to \prod(A) \le \prod(B);$$
$$\prod \left(\bigcup_{i \in I} A_i \right) = \sup_{i \in I} \prod(A_i) \tag{1.30}$$

is the possibility measure. Here I is index set.

In possibility theory every possibility measure is uniquely determined by the possibility distribution function.

Given a possibility measure $\prod$ on power set $P(X)$ of X, function $r : X \to [0, 1]$ such that

$$r(x) = \prod(\{x\})$$

for all $x \in X$ called a possibility distribution function of $\prod$.

We can write that

$$\prod(A) = \max_{x \in A} r(x)$$

for each $A \in P(X)$.

For $\forall A, B \subseteq P(X)$

$$\prod(A \cup B) = \max \left(\prod(A), \prod(B) \right)$$

When X is not finite, then

$$\prod(A) = \sup_{x \in A} r(x)$$

If A and $\overline{A}$ are two opposite events, then possibility measure satisfies

$$\max\left(\Pi(A), \Pi(\overline{A})\right) = 1 \quad \text{or} \quad \prod(A) + \prod(\overline{A}) \geq 1,$$

whereas for the probability measure $P()$ one has

$$P(A) + P(\overline{A}) = 1.$$

This represents the main difference between possibility and probability (Utkin 2007; Zadeh 1978). The probability value of an event is completely defined by the probability value of the opposite event. The possibility value of an event and the possibility value of the opposite one are weakly connected.

Possibility distribution function can be interpreted as a membership function of the fuzzy set; i.e. $r(x)$ is defined to be μ_A.

Possibility distribution is an important technique to represent imprecise information in the problem of approximate reasoning.

1.1.3   Type-1 Fuzzy Numbers

Definition 1.17   *Fuzzy number*. A type-1 fuzzy number is a fuzzy set A on R which possesses the following properties: (a) A is a normal fuzzy set; (b) A is a convex fuzzy set; (c) α-cut of A, A^{α} is a closed interval for every $\alpha \in (0, 1]$; (d) the support of A, A^{+0} is bounded.

In Fig. 1.2 some basic types of type-1 fuzzy numbers are shown. For comparison of a fuzzy number with a crisp number, in Fig. 1.3, the crisp number 2 is presented.

Intervals are special cases of type-1 fuzzy numbers. Interval analysis is frequently used in decision theories, in decision making with imprecise probabilities. Because of this first we will consider operations over intervals of real line. It is easy to see that a result of operations over closed intervals is also a closed interval. Due to this fact, the bounds of the resulting closed interval are to be expressed by means of the bounds of the given intervals.

Below we introduce the basic operations over closed intervals. Assume that the following intervals of the real line R are given:

$$A = [a_1, a_2]; \quad B = [b_1, b_2].$$

Addition of intervals. If $x \in [a_1, a_2]$ and $y \in [b_1, b_2]$, then

$$x + y \in [a_1 + b_1, a_2 + b_2] \tag{1.31}$$

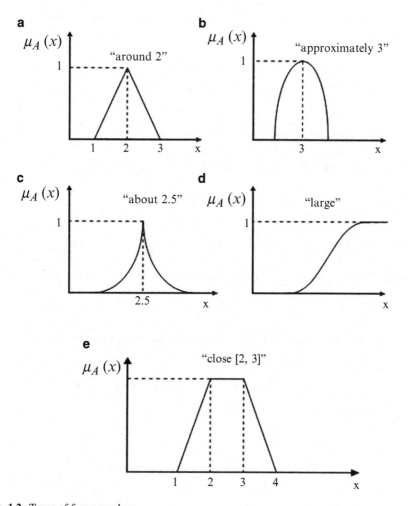

Fig. 1.2 Types of fuzzy numbers

Fig. 1.3 Crisp number 2

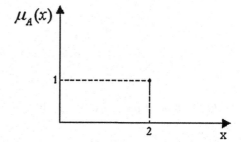

Fig. 1.4 Addition of closed
intervals

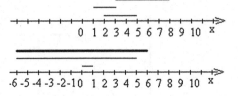

This can symbolically be expressed as (Fig. 1.4)

$$A + B = [a_1, a_2] + [b_1, b_2] = [a_1 + b_1, a_2 + b_2] \qquad (1.32)$$

Example

$$[2, 5] + [1.3] = [2 + 1, 5 + 3] = [3, 8]$$
$$[0, 1] + [-6, 5 = 0 - 6, 1 + 5 = -6, 6].$$

Image of interval A. If then $- x \in [-a_1, -a_2]$. This is symbolically expressed as

$$-A = [-a_2, -a_1]$$

Let us consider the result of $A + (-A)$. According to the operations given above,
we can write:

$$A + (-A) = [a_1, a_2] + [-a_2, -a_1] = [a_1 - a_2, a_2, a_1]$$

Note that $A + (-A) \neq 0$ (Fig. 1.5).

Example

$$-[1, 3] = [-3, -1];$$
$$[1, 3] + (-[1, 3]) = [1, 3] + [-3, -1] = [1 - 3, 3 - 1] = [-2, 2].$$

Subtraction of intervals. If $x \in [a_1, a_2]$ and $y \in [b_1, b_2]$, then (Fig. 1.6)

$$x - y \in [a_1 - b_2, a_2 - b_1] \qquad (1.33)$$
$$A - B = [a_1, a_2] - [b_1, b_2] = [a_1 - b_2, a_2 - b_1] \qquad (1.34)$$

Example

$$[2, 5] - [1, 3] = [2 - 3, 5 - 1] - [-1, 4],$$
$$[0, 1] - [-6, 5] = [0 - 5, 1 + 6] = [-5, 7].$$

Multiplication of intervals. The multiplication of intervals $A, B \subset R$ is defined as
follows:

$$A \cdot B = [\min(a_1 \cdot b_1, a_1 \cdot b_2, a_2 \cdot b_1, a_2 \cdot b_2), \max(a_1 \cdot b_1, a_1 \cdot b_1, a_2 \cdot b_1, a_2 \cdot b_2)] \quad (1.35)$$

Fig. 1.5 Addition of intervals A and $-A$

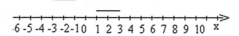

Fig. 1.6 Difference of intervals

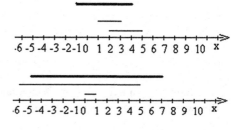

Fig. 1.7 Multiplication of intervals

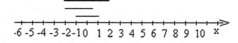

For the case $A, B \subset R^+$ the result is obtained easily as

$$A \cdot B = [a_1, a_2] \cdot [b_1, b_2] = [a_1 \cdot b_1, a_2 \cdot b_2] \qquad (1.36)$$

The multiplication of an interval A by a real number $k \in R$ is defined as follows:

if $k > 0$ then $k \cdot A = k \cdot [a_1, a_2] = [ka_1, ka_2]$,
if $k < 0$ then $k \cdot A = k \cdot [a_1, a_2] = [ka_2, ka_1]$ (Fig. 1.7).

Example

$$[-1, 1] \cdot [-2, 0.5] =$$
$$[\min((-1) \cdot (-2), (-1) \cdot 0.5, 1 \cdot (-2), 1 \cdot 0.5),$$
$$\max((-1) \cdot (-2), (-1) \cdot 0.5, 1 \cdot (-2), 1 \cdot 0.5) =$$
$$[\min(2, -0.5, -2, 0.5), \max(2, -0.5, -2, 0.5)] = [-2, 2].$$

Division of intervals. Under assumption that the dividing interval does not contain 0 and $A, B \subset R^+$ one has

$$A : B = [a_1, a_2] : [b_1, b_2] = [a_1/b_2, a_2 - b_1] \qquad (1.37)$$

Based on Eq. 1.37, the inverse of A can be defined as follows:
If $x \in [a_1, a_2]$ then $\frac{1}{x} \in \left[\frac{1}{a_2}, \frac{1}{a_1}\right]$
and

$$A^{-1} = [a_1, a_2]^{-1} = [1/a_2, 1/a_1] \qquad (1.38)$$

In general case the ratio of A and B can be done as follows

$$[a_1,a_2]:[b_1,b_2]=[a_1,a_2]\cdot[1/b_2,1.b_1]$$
$$=[\min(a_1/b_1,a_1/b_2,a_2/b_1,a_2/b_2),\max(a_1/b_1,a_1/b_2,a_2/b_1,a_2/b_2)] \tag{1.39}$$

Note that

$$A\cdot A^{-1}=[a_1/a_2,\,a_2/a_1]\neq 1$$

The division by a number $k>0$ is equivalent to multiplication by a number $1/k$. Below we list some important properties of operations over intervals.

1. **Commutativity**

$$A+B=B+A,$$
$$A\cdot B=B\cdot A$$

2. **Associativity**

$$A+(B+C)=(A+B)+C$$
$$(A\cdot B)\cdot C=A\cdot(B\cdot C)$$

3. **Idempotency**
   In case of $0\notin \mathrm{supp}(A)$ and $\mathrm{supp}(B\cdot C)>0$ one has

$$A+0=0+A=A$$
$$A\cdot 1=1\cdot A=A$$

4. **Distributivity**

$$A\cdot(B+C)=A\cdot B+A\cdot C$$

(distributivity property is partially satisfied).

Now we consider arithmetic operation on type-1 fuzzy numbers. There are different methods for developing type-1 fuzzy arithmetic. In this section we present these methods.

Method based on the extension principle. By this method basic arithmetic operations on real numbers are extended to operations on fuzzy numbers. Let A and B be two fuzzy numbers and $*$ denote any of four arithmetic operations $\{+,-,\cdot,:\}$.

A fuzzy set $A*B$ on R can be defined by the equation

$$\forall z\in \mathrm{R}\quad \mu_{(A*B)}(z)=\sup_{z=x*y}\min[\mu_A(x),\,\mu_B(y)] \tag{1.40}$$

It is shown in Klir and Yuan (1995) that $A * B$ is fuzzy number and the following theorem has been formulated and proved.

Theorem 1.1 *Let* $* \in \{+,-,\cdot,:\}$, *and let* A, B *denote continuous fuzzy numbers. Then, the fuzzy set* A * B *defined by Eq. 1.40 is a continuous fuzzy number. Then for four basic arithmetic operations on type-1 fuzzy numbers we can write*

$$\mu_{(A+B)}(z) = \sup_{z=x+y} \min[\mu_A(x), \mu_B(y)] \tag{1.41}$$

$$\mu_{(A-B)}(z) = \sup_{z=x-y} \min[\mu_A(x), \mu_B(y)] \tag{1.42}$$

$$\mu_{(A\cdot B)}(z) = \sup_{z=x\cdot y} \min[\mu_A(x), \mu_B(y)] \tag{1.43}$$

$$\mu_{(A:B)}(z) = \sup_{z=x:y} \min[\mu_A(x), \mu_B(y)] \tag{1.44}$$

Method based on interval arithmetic and α-cuts. This method is based on representation of arbitrary fuzzy numbers by their α-cuts and use interval arithmetic to the α-cuts. Let $A, B \subset R$ be fuzzy numbers and $*$ denote any of four operations. For each $\alpha \in (0, 1]$, the α-cut of $A * B$ is expressed as

$$(A*B)^\alpha = A^\alpha * B^\alpha \tag{1.45}$$

For $*$ we assume $0 \notin \text{supp}(B)$.
The resulting fuzzy number $A * B$ can be defined as

$$A*B = \bigcup_{\alpha \in [0,1]} \alpha(A*B)^\alpha \tag{1.46}$$

Next we using Eq. 1.45, Eq. 1.46 illustrate four arithmetic operations on fuzzy numbers.
Addition of fuzzy numbers. Let A and B be two fuzzy numbers and A^α and B^α their α cuts

$$A^\alpha = [a_1^\alpha, a_2^\alpha]; B^\alpha [b_1^\alpha, b_2^\alpha] \tag{1.47}$$

Then we can write

$$A^\alpha + B^\alpha = [a_1^\alpha, a_2^\alpha] + [b_1^\alpha, b_2^\alpha] = [a_1^\alpha + b_1^\alpha, a_2^\alpha + b_2^\alpha] \tag{1.48}$$

here

$$A^\alpha = \left\{ x/\mu\tilde{A}(x) \geq \alpha \right\}; B^\alpha = \left\{ x/\mu\tilde{B}(x) \geq \alpha \right\} \tag{1.49}$$

Subtraction of fuzzy numbers. Subtraction of given fuzzy numbers A and B can be defined as

$$(A - B)^\alpha = A^\alpha - B^\alpha = \left[a_1^\alpha - b_2^\alpha, a_2^\alpha - b_1^\alpha\right], \forall \alpha \in [0,1] \qquad (1.50)$$

We can determine Eq. 1.50 by addition of the image B^- to A

$$\forall \alpha \in [0,1], B^{\alpha^-} = \left[-b_2^\alpha, b_1^\alpha\right] \qquad (1.51)$$

Hukuhara difference. Hukuhara difference of given fuzzy numbers A and B can be defined as

$$(A -_h B)^\alpha = A^\alpha - B^\alpha = \left[a_1^\alpha - b_1^\alpha, a_2^\alpha - b_2^\alpha\right], \forall \alpha \in [0,1]$$

subject to $b_2^\alpha - b_1^\alpha \le a_2^\alpha - a_1^\alpha, \forall \alpha \in [0,1]$.
Multiplication of fuzzy numbers. Let two fuzzy numbers A and B be given. Multiplication $A \cdot B$ is defined as

$$(A \cdot B)^\alpha = A^\alpha \cdot B^\alpha = \left[a_1^\alpha, a_2^\alpha\right] \cdot \left[b_1^\alpha, b_2^\alpha\right] \forall \alpha \in [0,1] \qquad (1.52)$$

Multiplication of fuzzy number A in R by ordinary numbers $k \in R^+$ is performed as follows

$$\forall A \subset R \ \ kA^\alpha = \left[ka_1^\alpha, ka_2^\alpha\right] \qquad (1.53)$$

Division. Division of two fuzzy numbers A and B is defined by

$$A^\alpha : B^\alpha = \left[a_1^\alpha, a_2^\alpha\right] : \left[b_1^\alpha, b_2^\alpha\right] \forall \alpha \in [0,1] \qquad (1.54)$$

Definition 1.18 ***Absolute value of a type-1 fuzzy number.*** Absolute value of fuzzy number is defined as:

$$abs(A) = \begin{cases} \max(A, -A), & \text{for } R^+ \\ 0, & \text{for } R^- \end{cases} \qquad (1.55)$$

1.1.4   Fuzzy Relations. Linguistic Variables

In modeling systems the internal structure of a system must be described first. An internal structure is characterized by connections (associations) among the elements of system. As a rule these connections or associations are represented by means of relation. We will consider here type-1 fuzzy relations which gives us the representation about degree or strength of this connection.

There are several definitions of fuzzy relation (Kaufman 1973; Yeh and Bang 1975; Zadeh 1971). Each of them depends on various factors and expresses different aspects of modeling systems.

Definition 1.19 *Fuzzy relation.* Let $X_1, X_2, \ldots, X_n$ be nonempty crisp sets. Then, a $R(X_1, X_2, \ldots, X_n)$ is called a fuzzy relation of sets $X_1, X_2, \ldots, X_n$, if $R(X_1, X_2, \ldots, X_n)$ is the fuzzy subset given on Cartesian product $X_1 \times X_2 \times \ldots \times X_n$.

For $n = 2$ the fuzzy relation is called a binary fuzzy relation and is denoted as $R(X, Y)$. For three, four, or n sets the fuzzy relation is called ternary, quaternary, or n-ary, respectively.

In particular, if $X_1 = X_2 = \ldots = X_n = X$ we say that fuzzy relation R is given on set X among elements $x_1, x_2, \ldots, x_n \in X$.

Notice, that fuzzy relation can be defined in another way. Namely, by two ordered fuzzy sets.

Assume, two fuzzy sets $\mu_A(x)$ and $\mu_B(y)$ are given on crisp sets X and Y, respectively. Then, it is said, that fuzzy relation $R_{AB}(X, Y)$ is given on sets X and Y, if it is defined in the following way (Aliev 1995):

$$\mu_{R_{AB}}(x, y) = \min_{x, y} [\mu_A(x), \mu_B(y)]$$

for all pairs (x, y), where $x \in X$ and $y \in Y$. As above, fuzzy relation R_{AB} is defined on Cartesian product.

Let a type-1 fuzzy binary relation on set X be given. Consider the following three properties of relation R:

1. Fuzzy relation R is *reflexive*, if

$$\mu_R(x, x) = 1$$

   for all $x \in X$. If there exists $x \in X$ such that this condition is violated, then relation R is *irreflexive*, and if $R(x, x) = 0$ for all $x \in X$, the relation R is *antireflective*;

2. A fuzzy relation R is *symmetric* if it satisfies the following condition:

$$\mu_R(x, y) = \mu_R(y, x)$$

   for all $x, y \in X$. If from $R(x, y) > 0$ and $R(x, y) > 0$ follows $x = y$ for all $x, y \in X$ relation R is called *antisymmetric*;

3. A fuzzy relation R is *transitive* (or, more specifically, max-min transitive) if

$$\mu_R(x, z) \geq \max_{y \in Y} \min(\mu_R(x, y), \mu_R(y, z))$$

   is satisfied for all pairs $(x, z) \in X$.

Definition 1.20 *Fuzzy proximity.* A type-1 fuzzy relation is called a fuzzy proximity or fuzzy tolerance relation if it is reflexive and symmetric. A fuzzy relation is called a fuzzy similarity relation if it is reflexive, symmetric, and transitive.

Definition 1.21 *Fuzzy composition.* Let A and B be two fuzzy sets on $X \times Y$ and $Y \times Z$, respectively. A fuzzy relation R on $X \times Z$ is defined as

$$R = \left\{ \left((x,z), \mu_R(x,z)\right) \big| (x,z) \in X \times Z \right\} \tag{1.56}$$

Here

$$\mu_R : X \times Y \rightarrow [0,1]$$
$$(x,z) \mapsto \mu_R(x,z) = \mu_{A \circ B}(x,z) = \underset{y \in Y}{S}(T(\mu_A(x,y), \mu_B(y,z))) \tag{1.57}$$

For $x \in X$ and $z \in Z$ "T" and "S" are triangular norms and triangular conorms, respectively.

Definition 1.22 *Equivalence (similarity) relation.* If fuzzy relation R is reflexive, symmetric and transitive, then relation R is an equivalence relation or a similarity relation.

A fuzzy relation R is a fuzzy compatibility relation if it is reflexive and symmetric. This relation is cutworthy. Compatibility classes are defined by means of α-cut. In fact, using α-cut a class of compatibility relation is represented by means of crisp subset. Therefore a compatibility relation can also be represented by reflexive undirected graph.

Now consider fuzzy partial ordering. Let X be nonempty set. It is well known, that to order a set it is necessary to give an order relation on this set. But sometimes our knowledge and estimates of the elements of a set are not accurate and complete. Thus, to order such set the fuzzy order on set must be defined.

Definition 1.23 *Fuzzy partial ordering relation.* Let R be binary fuzzy relation on X. Then fuzzy relation R is called fuzzy partial ordering, if it satisfies the following conditions:

1. Fuzzy relation R is reflexive;
2. Fuzzy relation R is antisymmetric;
3. Fuzzy relation R is fuzzy transitive.

If fuzzy partial order is given on set X then we will say that set X is fuzzy partially ordered.

Next we consider projections and cylindrical extension.

Let R be n-dimensional fuzzy relation on Cartesian product $X = X_1 \times X_2 \times \ldots \times X_n$ of nonempty sets $X_1, X_2, \ldots, X_n$ and $(i_1, i_2, \ldots, i_k)$ be a sub-sequence of $(1, 2, \ldots, n)$.

The practice and experimental evidence have shown that decision theories developed for perfect decision-relevant information and 'well-defined' preferences are not capable of adequate modeling of real-world decision making. The reason is that real decision problems are characterized by imperfect decision-relevant information and vaguely defined preferences. This leads to the fact that when solving real-world decision problems we need to move away from traditional decision approaches based on exact modeling which is good rather for decision analysis of thought experiments.

More concretely, the necessity to sacrifice the precision and determinacy is by the fact that real-world problems are characterized by perception-based information and choices, for which natural language is more covinient and close than precise formal approaches. Modeling decision making from this perspective is impossible without dealing with fuzzy categories near to human notions and imaginations. In this connection, it is valuable to use the notion of linguistic variable first introduced by L. Zadeh (1975). Linguistic variables allow an adequate reflection of approximate in-word descriptions of objects and phenomena in the case if there is no any precise deterministic description. It should be noted as well that many fuzzy categories described linguistically even appear to be more informative than precise descriptions.

Definition 1.24 *Linguistic variable.* A linguistic variable is characterized by the set (u, T, X, G, M), where u is the name of variable; T denotes the term-set of u that refer to as base variable whose values range over a universe X; G is a syntactic rule (usually in form of a grammar) genera-ting linguistic terms; M is a semantic rule that assigns to each linguistic term its meaning, which is a fuzzy set on X.

A certain $t \in T$ generated by the syntactic rule G is called a term. A term consisting of one or more words, the words being always used together, is named an atomary term. A term consisting of several atomary terms is named a composite term. The concatenation of some components of a composite term (i.e. the result of linking the chains of components of the composite term) is called a subterm. Here $t_1, t_2, \ldots$ are terms in the following expression

$$T = t_1 + t_2 + \ldots$$

The meaning $M(t)$ of the term t is defined as a restriction $R(t; x)$ on the basis variable x conditioned by the fuzzy variable $\widetilde{X}$:

$$M(t) \equiv R(t; x)$$

It is assumed here that $R(t; x)$ and, consequently, $M(t)$ can be considered as a fuzzy subset of the set X named as t.

The assignment equation in case of linguistic variable takes the form in which t-terms in T are names generated by the grammar G, where the meaning assigned to the term t is expressed by the equality

$$M(t) = R(term \, in \, T)$$

In other words the meaning of the term t is found by the application of the semantic rule M to the value of term t assigned according to the right part of equation. Moreover, it follows that $M(t)$ is identical to the restriction associated with the term t.

It should be noted that the number of elements in T can be unlimited and then for both generating elements of the set T and for calculating their meaning, the application of the algorithm, not simply the procedure for watching term-set, is necessary.

We will say that a linguistic variable u is structured if its term-set T and the function M, which maps each element from the term-set into its meaning, can be given by means of algorithm. Then both syntactic and semantic rules connected with the structured linguistic variable can be considered algorithmic procedures for generating elements of the set T and calculating the meaning of each term in T, respectively.

However in practice we often encounter term-sets consisting of a small number of terms. This makes it easier to list the elements of term-set T and establishes a direct mapping from each element to its meaning. For example, an intuitive description of possible economic conditions may be represented by linguistic terms like "strong economic growth", "weak economic growth" etc. Then the term set of linguistic variable "state of economy" can be written as follows:

$$T(\text{state of economy}) = \text{"strong growth"} + \text{"moderate growth"} + \text{"static situation"} + \text{"recession"}.$$

The variety of economic conditions may also be described by ranges of the important economic indicators. However, numerical values of indicators may not be sufficiently clear even for experts and generate questions and doubts. In contrast, linguistic description is well perceived by human intuition as qualitative and fuzzy.

1.1.5   Type-1 Fuzzy Logic

We will consider the logics with multi-valued and continuous values. Let's define the semantic truth function of this logic. Let P be statement and $T(P)$ its truth value, where $T(P) \in [0, 1]$

Negation values of the statement P are defined as:

$$T(\neg P) = 1 - T(P)$$

Hence

$$T(\neg\neg P) = T(P)$$

The implication connective is always defined as follows:

$$T(P \rightarrow Q) = T(\neg P \vee Q)$$

and the equivalence as

$$T(P \leftrightarrow Q) = T[(P \rightarrow Q) \wedge (Q \rightarrow P)]$$

It should be noted that exclusive disjunction ex, disjunction of negations (Shiffer's connective) |, conjunction of negations ↓ and $\sim \rightarrow$ (has no common name) are defined as negation of equivalence $\leftrightarrow$, con-junction $\wedge$, disjunction $\vee$, and implication $\rightarrow$, respectively.

The tautology denoted • and contradiction denoted ° will be, respectively:

$$T\left(\overset{\bullet}{P}\right) = T(P \vee \neg P); T\left(\overset{\circ}{P}\right) = T(P \wedge \neg P)$$

More generally

$$T\left(\overset{\bullet}{P}Q\right) = T((P \vee \neg P) \vee (Q \vee Q))$$

$$T\left(\overset{\circ}{P}Q\right) = T((P \wedge \neg P) \wedge (Q \wedge Q))$$

Let us define the basic connectives of fuzzy logic in the following two fuzzy set theories.

Logic based on $(P(X), \cap, \cup, -)$. In this case disjunction and conjunction are defined as:

$$T(P \vee Q) = \max(T(P), T(Q)); T(P \wedge Q) = \min(T(P), T(Q))$$

It is clear, that $\vee$ and $\wedge$ are commutative, associative, idempotent and distributive and do not satisfy the law of excluded-middle, i.e. $T(P \vee \neg P) \neq 1$ and $T(P \vee \neg P) \neq 0$, but satisfy absorption law

$$T(P \vee (P \wedge Q)) = T(P); T(P \wedge (P \vee Q))T(P)$$

and also *De Morgan's laws:*

$$T(\neg(P \wedge Q)) = T(\neg P \vee \neg Q)$$
$$T(\neg(P \vee Q)) = T(\neg P \vee \neg Q)$$

Equivalence is defined as

$$T[(\neg P \vee Q) \wedge (P \vee \neg Q)] = T(\neg P \wedge Q) \vee (\neg P \vee \neg Q)$$

Law of excluded disjunction:

$$T[(\neg P \wedge Q) \vee (P \wedge \neg Q)] = [T(P \vee Q) \wedge (\neg P \vee \neg Q)]$$

The expressions for 16 connectives are presented in Table 1.2. It is assumed here that $T(P) = P$ and $T(Q) = q$.

The quantifiers in the statements will be:

$$T(\exists x P(x)) = \sup(T(P(x))); \; T(\forall x P(x)) = \inf(T(P(x)))$$

where x denotes an element of the universe of discourse.

Multi-valued logic based on $\left(\widetilde{P}(X), \cap, \cup, -\right)$ usually is called K- standard sequence logic. In this logic the connectives satisfy the following properties:

Implication:

$$T[P \to (Q \to R)] = T[(P \wedge Q) \to R];$$

Tautology and contradiction:

$$T(P \to P) = T\left(\overset{\bullet}{P}\right)$$

$$T\left(\overset{\circ}{P} \to P\right) = T(P)$$

$$T\left(P \to \overset{\circ}{P}\right) = T\left(\overset{\bullet}{P}\right)$$

$$T(P \leftrightarrow P) = T\left(\overset{\bullet}{P}\right)$$

$$T\left(\overset{\circ}{P} \to P\right) = T\left(\overset{\bullet}{P}\right)$$

$$T\left(P \to \overset{\circ}{P}\right) = T(\neg P)$$

$$T(P \leftrightarrow \neg P) = T\left(\overset{\circ}{P}\right)$$

The Shiffer's and Pierce's connectives:

$$T(\neg P) = T(P|P)$$

$$T(P \to Q) = T(P|(Q/Q))$$

$$T\left(\overset{\circ}{P}\right) = T(P|(P|P))$$

It is shown in Dubois and Prade (1980), Klir and Yuan (1995) that the multi-valued logic is fuzzification of the traditional propositional calculus (in sense of

Table 1.2 Expressions for connectives

PQ	$P \overset{\bullet}{\vee} Q$	$P \vee Q$	$Q \to P$	P
	$\max(p, 1-p, q, 1-q)$ $Q \to P$	$\max(p, q)$ Q	$\max(p, 1-q)$ $Q \leftrightarrow P$	p $P \wedge Q$
	$\max(1-p, q)$ $P\vert Q$	q $PexQ$	$\min\big(\max(1-p, q),$ $\max(p, 1-q)\big)$ $\neg Q$	$\min(p, q)$ $Q - \to$
	$\max(1-p, 1-q)$ $\neg P$	$\max\big(\min(1-p, q),$ $pq(p, 1-q)\big)$ $P - \to Q$	$1-q$ $P \downarrow Q$	$\min(p, 1-q)$ $P \overset{\bullet}{\vee} Q$
	$\min(1-p, 1-q)$	$\min(1-p, q)$	$\min(1-p, 1-q)$	$\min(p, 1-q,$ $q, 1-q)$

extension principle). In this logic each proposition P is assigned a normalized type-1 fuzzy set in [0,1], i.e. the pair $\{\mu_p(0), \mu_p(1)\}$ is interpreted as the degree of false or truth, respectively. Because the logical connectives of standard propositional calculus are functionals of truthness, that is they are represented as functions, they can be fuzzified.

Logic based on $\left(\widetilde{P}(X), \cap_{\bullet}, \cup^{\bullet}, -\right)$. In this case disjunction and conjunction are defined as

$$T\left(P \overset{\bullet}{\vee} Q\right) = \min(1, T(P) + T(Q))$$

$$T\left(P \underset{\bullet}{\wedge} Q\right) = \max(0, T(P) + T(Q) - 1)$$

It is clear that $\vee^{\bullet}$ and $\wedge_{\bullet}$ are commutative, associative, not idempotent and not distributive and they satisfy De Morgan's laws

$$T\left(\neg\left(P \overset{\bullet}{\vee} Q\right)\right) = T(\neg P \vee \neg Q)$$

$$T\left(\neg\left(P \overset{\bullet}{\wedge} Q\right)\right) = T(\neg P \wedge \neg Q)$$

and the law of excluded-middle

$$T(P \vee \neg P) = 1, T(P \wedge \neg P) = 0$$

The 16 connectives are given in the Table 1.3. Here $\vee, \to, \leftrightarrow, \wedge, \vert, ex, \sim\to, \downarrow$ are denoted by $\vee^{\bullet}, \Rightarrow, \Leftrightarrow, \wedge_{\bullet}, \Vert, ex, \approx\Rightarrow, \downarrow\downarrow$, respectively.

Tautology and contradiction satisfy the following properties:

$$T(P \Rightarrow P) = T\left(\overset{\bullet}{P}\right)$$
$$T\left(\overset{\bullet}{P} \Rightarrow P\right) = T\left(\overset{\bullet}{P}\right)$$
$$T\left(P \Rightarrow \overset{\bullet}{P}\right) = T\left(\overset{\bullet}{P}\right)$$
$$T(P \Leftrightarrow P) = T\left(\overset{\bullet}{P}\right)$$
$$T\left(\overset{\circ}{P} \Rightarrow P\right) = T\left(\overset{\bullet}{P}\right)$$
$$T\left(P \Leftrightarrow \overset{\circ}{P}\right) = T(\neg P)$$

In Zadeh's notation the implication $\Rightarrow$ corresponds to the usual inclu-sion for fuzzy sets, ex and $\approx \Rightarrow$ correspond to symmetric ∇ and bounded $|-|$ differences, respectively. This logic is known as Lukasiewicz logic (L-logic).

It should be noted that these two theories of type-1 fuzzy sets and logics constructed on the basis of these theories are not only known at the present time. In connection with this it is necessary to give semantic analysis of the major known multi-valued logics. For this purpose we will use power sets which are necessary for formalization of some operations on fuzzy sets.

Semantic analysis of different type-1 fuzzy logics. Let A and B be fuzzy sets of the subsets of non-fuzzy universe U; in fuzzy set theory it is known that $\widetilde{A}$ is a subset of B iff

$$\mu_A \leq \mu_B, \text{ i.e. } \forall x \in U, \ \mu_A(x) \leq \mu_B(x).$$

Definition 1.25 *Power fuzzy set.* For given fuzzy implication $\rightarrow$ and fuzzy set B from the universe U, the power fuzzy set PB from B is given by membership function μ_{PB} (Bandler and Kohout 1980):

$$\mu_{PB}A \underset{x \in U}{\wedge} (\mu_A(x) \rightarrow \mu_B(x))$$

Then the degree to which A is subset of B, is

$$\pi(A \subseteq B) = \mu_{PB}A$$

Definition 1.26 *Fuzzy implication operator.* Given fuzzy implication operator $\rightarrow$ on the closed unit interval [0,1] then:

$$a \leftarrow b = b \rightarrow a$$
$$a \leftrightarrow b = (a \rightarrow b) \wedge (a \leftarrow b) = (a \rightarrow b) \wedge (a \leftarrow b)$$

Definition 1.27 *Degree of equivalency.* Under the conditions of the definition PB the degree to which fuzzy sets A and B are equivalent is:

Table 1.3 Expressions for connectives

PQ	$\dot{P}Q$	$Q \Rightarrow P$	$Q \Rightarrow P$	P
pq PQ	1 $Q \Rightarrow P$	$\min(1, p+q)$ Q	$\min(1, p+1-q)$ $P \Leftrightarrow Q$	p $P \wedge Q$
pq PQ	$\min(1, 1-p+q)$ $P\|Q$	Q $PexQ$	$1-\|p-q\|$ $\neg Q$	$\max(0, p+q-1)$ $Q \approx\Rightarrow P$
pq PQ	$\min(1, 1-p+1q)$ $\neg P\|$	$\|p-q\|$ $P \approx\Rightarrow Q$	$1-q$ $P \downarrow\downarrow Q$	$\max(0, p-q)$ $P \overset{\circ}{} Q$
pq	$1-p$	$\max(0, q-p)$	$\max(0, 1-p-q)$	0

$$\pi(A \equiv B) = \pi(A \subseteq B) \wedge \pi(B \subseteq A)$$

or

$$\pi(A \equiv B) = \bigwedge_{x \in U} \left(\mu_A x \rightarrow \mu_B x \right)$$

For practical purposes (Bandler and Kohout 1980) in most cases it is advisable to work with multi-valued logics in which logical variable takes values from the real interval $I = [0, 1]$ divided into ten subintervals, i.e. by using set $V_{11} = [0, 0.1, 0.2, \ldots, 1]$.

We denote the truth values of premises A and B through $T(A) = a$ and $T(B) = b$. The implication operation in analyzed logics (Aliev 1995) has the following form:

1. **Min-logic**

$$a \underset{\min}{\rightarrow} b = \begin{cases} a, \text{ if } a \leq b \\ b, \text{ otherwise.} \end{cases}$$

2. **$S^{\#}$-logic**

$$a \underset{S^{\#}}{\rightarrow} b = \begin{cases} 1, \text{ if } a \neq 1 \text{ or } b = 1, \\ 0, \text{ otherwise.} \end{cases}$$

3. **S-logic (Standard sequence)**

$$a \underset{S}{\rightarrow} b = \begin{cases} 1, \text{ if } a \leq b, \\ 0, \text{ otherwise.} \end{cases}$$

4. **G-logic (Gödelian sequence)**

$$a \underset{G}{\rightarrow} b = \begin{cases} 1, \text{ if } a \leq b, \\ b, \text{ otherwise.} \end{cases}$$

5. **G43-logic**

$$a \underset{G43}{\rightarrow} b = \begin{cases} 1, \text{ if } a = 0, \\ \min(1, b/a), \text{ otherwise.} \end{cases}$$

6. **L-logic (Lukasiewicz's logic)**

$$a \underset{L}{\rightarrow} b = \min(1, 1 - a + b)$$

7. **KD-logic**

$$a \underset{KD}{\rightarrow} b = ((1 - a) \vee b = \max(1 - a, b)$$

In turn *ALI1-ALI4* – logics, suggested by us, are characterized by the following implication operations (Aliev et al. 1991b; Aliev and Aliev 1997–1998):

8. **ALI1-logic**

$$a \underset{ALI1}{\rightarrow} b = \begin{cases} 1 - a, \text{ if } a < b, \\ 1, \text{ if } a = b, \\ b, \text{ if } a > b \end{cases}$$

9. **ALI2-logic**

$$a \underset{ALI2}{\rightarrow} b = \begin{cases} 1, \text{ if } a \leq b, \\ (1 - a) \wedge b, \text{ if } a > b \end{cases}$$

10. **ALI3-logic**

$$a \underset{ALI3}{\rightarrow} b = \begin{cases} 1, \text{ if } a \leq b, \\ b/[a + (1 - b)], \text{ otherwise.} \end{cases}$$

11. **ALI4-logic**

$$a \underset{ALI4}{\rightarrow} b = \begin{cases} \dfrac{1 - a + b}{2}, \text{ } a < b, \\ 1, \text{ } a \leq b. \end{cases}$$

A necessary observation to be made in the context of this discussion is that with the only few exceptions for *S-logic* (3) and *G-logic* (4), and *ALI1-ALI4* (8)–(11), all other known type-1 fuzzy logics (1)–(2), (5)–(7) do not satisfy either the classical "modus-ponens" principle, or other criteria which appeal to the human perception of mechanisms of a decision making process being formulated in Mizumoto et al. (1979). The proposed type-1 fuzzy logics *ALI1-ALI4* come with an implication operators, which satisfy the classical principle of "modus-ponens" and meets some additional criteria being in line with human intuition.

The comparative analysis of the first seven logics has been given in Bandler and Kohout (1980). The analysis of these seven logics has shown that only *S*- and *G*-logics satisfy the classical principle of Modus Ponens and allow development of improved rule of fuzzy conditional inference. At the same time the value of truthness of the implication operation in *G*-logic is equal either to 0 or 1; and only the value of truthness of logical conclusion is used in the definition of the implication operation in *S*-logic. Thus the degree of "fuzziness" of implication is decreased, which is a considerable disadvantage and restricts the use of these logics in approximate reasoning.

Definition 1.28 *Top of a fuzzy set.* The top of fuzzy set. *B* is

$$HB = \bigvee_U \mu_B(x).$$

Definition 1.29 *Bottom of a fuzzy set.* The bottom of fuzzy set *B* is

$$pB = \bigwedge_U \mu_B(x).$$

Definition 1.30 *Nonfuzziness.* Nonfuzziness $a \in U$ is $ka = a \vee (1-a)$. Then nonfuzziness of fuzzy set *B* is defined as:

$$kB = \bigwedge_U k\mu_B(x)$$

Let us give a brief semantic analysis of the proposed fuzzy logics *ALI1-ALI3* by using the terminology accepted in the theory of power fuzzy sets. For this purpose we formulate the following.

Proposal. Possibility degree of the inclusion of set $\pi(A \subseteq B)$ in fuzzy logic *ALI1-ALI3* is determined as:

$$\pi_1(A \subseteq B) = \begin{cases} 1 - \mu_A(x), & \text{if } \mu_A(x) < \mu_B(x), \\ 1, & \text{if } \mu_A(x) = \mu_B(x), \\ \mu_B(x), & \text{if } \mu_A(x) > \mu_B(x); \end{cases}$$

$$\pi_2(A \subseteq B) = \begin{cases} 1, & \text{if } \mu_A(x) \le \mu_B(x), \\ (1 - \mu_A(x)) \wedge \mu_B(x), & \text{if } \mu_A(x) > \mu_B(x); \end{cases}$$

$$\pi_3(A \subseteq B) = \begin{cases} 1, \text{ if } \mu_A(x) \le \mu_B(x), \\ \dfrac{\mu_B(x)}{\mu_A(x) + (1 - \mu_B(x))}, \text{ if } \mu_A(x) > \mu_B(x). \end{cases}$$

We note, that if $\mu_A(x) = 0$ or $A \ne \emptyset$, then the crisp inclusion is possible for fuzzy logic *ALI1*. Below we consider the equivalence of fuzzy sets.

Proposal. Possibility degree of the equivalence of the sets $\pi(A \equiv B)$ is determined as:

$$\pi_1(A \equiv B) = \begin{cases} 1 - [(1 - \mu_A(x)) \vee \mu_B(x)], \text{ if } \mu_A(x) < \mu_B(x), \\ 1, \text{ if } A = B, \\ 1 - [(1 - \mu_B(x)) \vee \mu_A(x)], \text{ if } \mu_A(x) > \mu_B(x), \end{cases}$$

$$\pi_2(A \equiv B) = \begin{cases} 1, \text{ if } A = B, \\ \bigwedge_T \{[(1 - \mu_A(x)) \wedge \mu_B(x)], [(1 - \mu_B(x)) \wedge \mu_A(x)]\} \text{ if } A \ne B, \\ 0, \text{ if } \exists x \mid\mid\mid \mu_A(x) = 0, \ \mu_B(x) \ne 0 \text{ (or vice versa)}, \\ \text{and also } \exists x \mid\mid\mid \mu_A(x) = 1, \ \mu_B(x) \ne 1 \text{(or vice versa)}, \end{cases}$$

$$\pi_3(A \subseteq B) = \begin{cases} 1, \text{ if } \mu_A(x) \le \mu_B(x), \\ \dfrac{\mu_B(x)}{\mu_A(x) + (1 - \mu_B(x))}, \text{ if } \mu_A(x) > \mu_B(x). \end{cases}$$

Here the set $T = \{x \in U | \mu_{Ax} \ne \mu_{Bx}\}$ and $A = B$ means that $\forall x\, \mu_A(x) = \mu_B(x)$ or in other words, $T = \emptyset$.

The symbol $\mid\mid\mid$ means "such as". From the expression $\pi_i(A \equiv B)$, $i = 1, 2, 3$, it follows that for *ALI1* fuzzy logic the equivalency $\pi_1(A \equiv B) = 1$ takes place only when $A = B$. It is obvious that the equivalence possibility is equal to 0 only in those cases when one of the statements is crisp, i.e. either true or false, while the other is fuzzy.

Proposal. Degree to which fuzzy set B is empty $\pi(B \equiv \emptyset)$ is determined as

$$\pi_1(B \equiv \emptyset) = \begin{cases} 1, \text{ if } B = \emptyset, \\ 0, \text{ otherwise}; \end{cases}$$

$$\pi_2(B \equiv \emptyset) = \begin{cases} 1, \text{ if } HB < 1 \text{ or } B = \emptyset, \\ 0, \text{ otherwise}; \end{cases}$$

$$\pi_3(B \equiv \emptyset) = \begin{cases} 1, \text{ if } B = \emptyset, \\ 0, \text{ otherwise}. \end{cases}$$

Here $B = \emptyset$ means that for $\forall x\, \mu_B(x) = 0$, or equivalently $HB = 0$.

We introduce the concept of disjointness of fuzzy sets. There are two kinds of the disjointness. For a set A the first kind is defined by degree to which set A is a subset of the complement of B^c. The second kind is the degree to which the intersection of sets is empty. Therefore, we formulate the following.

Proposal. Degree of disjointness of sets A and B is degree to which A and B are disjoint

$$\pi(A \ disj_1 \ B) = \pi(A \subseteq B) \wedge \pi(B \subseteq A),$$
$$\pi(A \ disj_2 \ B) = \pi((A \cap B) = \varnothing).$$

Proposal. Disjointness grade of sets A and B is determined as

$$\pi_1(A \ disj_1 \ B) = \begin{cases} 1, & \text{if } \exists x \big|\big|\big| \mu_A(x) = 1 - \mu_B(x), \\ (1 - \mu_A(x)) \wedge (1 - \mu_B(x)), & \text{otherwise,} \\ 0, & \text{never;} \end{cases}$$

$$\pi_2(A \ disj_1 \ B) = \begin{cases} 1, & \text{if } \mu_A(x) \leq 1 - \mu_B(x), \\ 0, & \text{if } \exists x \big|\big|\big| \mu_A(x) = 1 \text{ and } \mu_B(x) \neq 0, \\ & \text{or } \mu_B(x) = 1 \text{ and } \mu_A(x) \neq 0, \\ \underset{T}{\wedge} [(1 - \mu_A(x)), (1 - \mu_B(x))], & \text{otherwise;} \end{cases}$$

$$\pi_3(A \ disj_1 \ B) = \begin{cases} 1, & \text{if } \mu_A(x) = \mu_B(x) \text{ or } \mu_B(x) = 0, \\ \underset{T}{\wedge} \left[\dfrac{1 - \mu_B(x)}{\mu_A(x) + (1 - \mu_B(x))}, \dfrac{1 - \mu_A(x)}{\mu_B(x) + (1 - \mu_A(x))} \right], & \text{otherwise,} \\ 0, & \text{never.} \end{cases}$$

Here $T = \{x \mid \mid \mid \mu_A(x) > 1 - \mu_B(x)\}$

We note that, the disjointness degree of the set is equal to 0 only for fuzzy logic ALI2, when under the condition that one of the considered fuzzy sets is normal, the other is subnormal.

Proposal. Degree to which set is a subset of its complement for the considered fuzzy logics $\pi_i(A \subseteq B^C)$ takes the following form

$$\pi_1(A \subseteq A^C) = \begin{cases} 1, & \text{if } HA = 0, \\ 0, & \text{if } HA = 1, \\ 1 - HA, & \text{otherwise;} \end{cases}$$

$$\pi_2(A \subseteq A^C) = \begin{cases} 1, & \text{if } HA \leq 0, \\ 0, & \text{if } HA = 1, \\ 1 - HA, & \text{otherwise;} \end{cases}$$

$$\pi_3(A \subseteq A^C) = \begin{cases} 1, & \text{if } HA \leq 0.5, \\ 0, & \text{if } HA = 1, \\ (1 - HA)/(2HA), & \text{otherwise;} \end{cases}$$

It is obvious that for the fuzzy logic *ALI1* the degree to which a set is the subset of its complement is equal to the degree to which this set is empty. It should also be mentioned that the semantic analysis given in Aliev et al. (1991b), Aliev and Aliev (1997–1998), Aliev et al. (1993) as well as the analysis given above show a significant analogy between features of fuzzy logics *ALI1* and *KD*. However, the fuzzy logic *ALI1*, unlike the *KD* logic, has a number of advantages. For example, *ALI1* logic satisfies the condition $\mu_A x \wedge (\mu_A x \rightarrow \mu_B x) \leq \mu_B x$ necessary for development of fuzzy conditional inference rules. *ALI2* and *ALI3* logics satisfy this inequality as well. This allows them to be used for the formalization of improved rules of fuzzy conditional inference and for the modeling of relations between main elements of a decision problem under uncertainty and interaction among behavioral factors.

1.1.6 Approximate Reasoning

In our daily life we often make inferences where *antecedents* and *consequents* are represented by fuzzy sets. Such inferences cannot be realized adequately by the methods, which are based either on two-valued logic or many-valued logic. In order to facilitate such an inference, Zadeh (1965, 1973, 1975, 1988, 2005, 2008) suggested an inference rule called a "compositional rule of inference". Using this inference rule, Zadeh, Mamdani (1977), Mizumoto (Fukami et al. 1980; Mizumoto et al. 1979; Mizumoto and Zimmermann 1982), R.Aliev and A.Tserkovny (Aliev et al. 2004; Aliev et al. 1991a; Aliev and Tserkovny 1988) suggested several methods for fuzzy reasoning in which the antecedent contain a conditional proposition involving fuzzy concepts:

$$
\begin{array}{l}
\text{Ant 1} : \text{ IF } x \text{ is } P \text{ THEN } y \text{ is } Q \\
\text{Ant 2} : \quad x \text{ is } P'
\end{array}
$$

$$\text{------------------------------------} \tag{1.58}$$

$$\text{Cons} : \text{ y is } Q'.$$

Those methods are based on implication operators present in various type-1 fuzzy logics. We will begin in fuzzy sets research the great attention is paid to the development of Fuzzy Conditional Inference Rules (CIR) (Aliev and Aliev 1997–1998; Fan and Feng 2009; Li et al. 2010; Medina and Ojeda-Aciego 2010). This is connected with the feature of the natural language to contain a certain number of fuzzy concepts (*F*-concepts), therefore we have to make logical inference in which the preconditions and conclusions contain such F-concepts. The practice shows that there is a huge variety of ways in which the formalization of rules for such kind of inferences can be made. However, such inferences cannot be satisfactorily formalized using the classical Boolean Logic, i.e. here we need to use multi-valued logical systems. The development of the conditional logic rules embraces mainly three types of fuzzy propositions:

$$P_1 = \text{IF} x \text{ is } A \text{ THEN} y \text{ is} B$$

$$P_2 = \text{IF} x \text{ is } A \text{ THEN} y \text{ is } B \text{ OTHERWISE } C$$
$$P_3 = \text{IF} x_1 \text{ is} A_1 \text{ and} x_2 \text{ is} A_2 \ldots \text{ and } \ldots \text{ and } x_n \text{ is} A_n \text{ THEN} y \text{ is} B$$

The conceptual principle in the formalization of fuzzy rules is the Modus Ponens inference (separation) rule that states:

IF $\alpha \rightarrow \beta$ is true and α is true THEN β is true.

The methodological base for this formalization is the compositional rule suggested by L.Zadeh (Zadeh 1965). Using this rule, he formulated some inference rules in which both the logical preconditions and consequences are conditional propositions including F-concepts. Later E.Mamdani (Mamdani 1977) suggested inference rule, which like Zadeh's rule was developed for the logical proposition of type P_1. In other words the following type F-conditional inference is considered:

$$
\begin{array}{ll}
\text{Proposition 1}: & \text{IF } x \text{ is } A \text{ THEN } y \text{ is } B \\
\text{Proposition 2}: & x \text{ is } A' \\
\hline
\text{Conclusion}: & y \text{ is } B'.
\end{array}
\tag{1.59}
$$

where A and A' are F-concepts represented as type-1 F-sets in the universe U; B is F-conceptions or F-set in the universe V. It follows that B' is the consequence represented as a F-set in V. To obtain a logical conclusion based on the CIR, the Propositions 1 and 2 must be transformed accordingly to the form of binary F-relation $R(A_1(x)), A_2(y))$ and unary F-relation $R(A_1(x))$. Here $A_1(x)$ and $A_2(y)$ are defined by the attributes x and y which take values from the universes U and V, respectively. Then

$$R(A_1(x)) = A' \tag{1.60}$$

According to Zadeh-Mamdani's inference rule $R(A_1(x)), A_2(y))$ is defined as follows.

The maximin conditional inference rule:

$$R_m(A_1(x), A_2(y)) = (A \times B) \cup (\neg A \times V) \tag{1.61}$$

The arithmetic conditional inference rule:

$$R_a(A_1(x), A_2(y)) = (\neg A \times V) \oplus (U \times B) \tag{1.62}$$

The mini-functional conditional inference rule:

$$R_c(A_1(x), A_2(y)) = A \times B \tag{1.63}$$

where $\times$, $\cup$ and $\neg$ are the Cartesian product, union, and complement operations, respectively; $\oplus$ is the limited summation.

Thus, in accordance with (Mamdani 1977; Zadeh 1965), the logical consequence $R(A_2(y))$ (B' in Eq. 1.59) can be derived as follows:

$$R(A_2(y)) = A' \circ [(A \times B)] \cup [\neg A \times U)]$$
$$R(A_2(y)) = A' \circ [(\neg A \times V)] \oplus [\neg U \times B)]$$

or

$$R(A_2(y)) = A' \circ (A \times B)$$

where $\circ -$ is the F-set maximin composition operator.

On the base of these rules the conditional inference rules for type P_2 were suggested in Baldwin and Pilsworth (1979):

$$\begin{aligned} &R_4(A_1(x), A_2(y)) \\ &= [(A \times V) \oplus (U \times B)] \cap [(A \times V) \oplus (U \times C)] \end{aligned} \qquad (1.64)$$

$$\begin{aligned} &R_5(A_1(x), A_2(y)) \\ &= [(\neg A \times V) \cup (U \times B)] \cap [(A \times V) \cup (U \times C)] \end{aligned} \qquad (1.65)$$

$$R_6(A_1(x), A_2(y)) = [(A \times B) \cup (\neg A \times C)] \qquad (1.66)$$

Note that in Baldwin and Pilsworth (1979) also the fuzzy conditional inference rules for type P_3 were suggested:

$$R_7(A_1(x), A_2(y)) = \left[\bigcap_{i=1,n} (\neg A_i \times V) \right] \oplus [(U \times B)] \qquad (1.67)$$

$$R_8(A_1(x), A_2(y)) = \left[\bigcap_{i=1,n} (\neg A_i \times V) \right] \cup [(U \times B)] \qquad (1.68)$$

$$\begin{aligned} R_9(A_1(x), A_2(y)) &= (\neg A \times V) \oplus (U \times B) \\ &= \int_{U \times V} 1 \wedge (1 - \mu_A(u) + \mu_B(v))/(u, v) \end{aligned} \qquad (1.69)$$

In order to analyze the effectiveness of rules (1.59), (1.60), (1.61), (1.62), (1.63), (1.64), (1.65), (1.66), (1.67), (1.68), and (1.69) we use some criteria for F-conditional logical inference suggested in Fukami et al. (1980). The idea of these criteria is to compare the degree of compatibility of some fuzzy conditional inference rules with the human intuition when making approximate reasoning. These criteria are the following:

Criterion I	Precondition 1: IF x is A THEN y is B Precondition 2: x is A
	Conclusion: y is B
Criterion II-1	Precondition 1: IF x is A THEN y is B Precondition 2: x is very A
	Conclusion: y is very B
Criterion II-2	Precondition 1: IF x is A THEN y is B Precondition 2: x is very A
	Conclusion: y is B
Criterion III	Precondition 1: IF x is A THEN y is B Precondition 2: x is more or less A
	Conclusion: y is more or less B
Criterion IV-1	Precondition 1: IF x is A THEN y is B Precondition 2: x is not A
	Conclusion: y is unknown
Criterion IV-2	Precondition 1: IF x is A THEN y is B Precondition 2: x is not A
	Conclusion: y is not B

In Fukami et al. (1980) it was shown that in Zadeh-Mamdani's rules the relations R_m, R_c and R_c do not always satisfy the above criteria. For the case of mini-operational rule R_c it has been found that criteria I and II-2 are satisfied while criteria II-1 and III are not.

In Fukami et al. (1980) an important generalization was made that allows some improvement to the mentioned F-conditional logical inference rules. It was shown there that for the conditional proposition arithmetical rule defined by Zadeh

$$P_1 = \text{IF} x \text{ is} A \text{ THEN} y \text{ is} B$$

the following takes place

$$R_9(A_1(x), A_2(y)) = (\neg A \times V) \oplus (U \times B)$$
$$= \int_{U \times V} 1 \wedge (1 - \mu_A(u) + \mu_B(v))/(u, v)$$

The membership function for this F-relation is

$$1 \wedge (1 - \mu_A(u) + \mu_B(v))$$

that obviously meets the implication operation or the Ply-operator for the multi-valued logic L (by Lukasiewicz), i.e.

$$T\left(P \underset{L}{\to} Q\right), T(P) \tag{1.70}$$

where $T(P \to_L Q), T(P)$ and $T(Q)$ are the truth values for the logical propositions $P \to_L Q, P$ and Q respectively.

In other words, these expressions can be considered as adaptations of implication in the L-logical system to a conditional proposition.

Having considered this fact, the following expression was derived:

$$
\begin{aligned}
R_a(A_1(x), A_2(y)) &= (\neg A \times V) \oplus (U \times B) \\
&= \int\limits_{U \times V} 1 \wedge (1 - \mu_A(u) + \mu_B(v))/(u, v) \\
&= \int\limits_{U \in V} \left(\mu_A(u) \underset{L}{\rightarrow} \mu_B(v)\right)/(u, v) = (A \times V) \rightarrow (U \times B)
\end{aligned}
\tag{1.71}
$$

In Fukami et al. (1980) an opinion was expressed that the implication operation or the Ply-operator in the expression (1.71) may belong to any multi-valued logical system. The following are guidelines for deciding which logical system to select for developing F-conditional logical inference rules (Fukami et al. 1980). Let F-sets A from U and B from V are given in the form:

$$
A = \int\limits_V \mu_A(u)/u, \ B = \int\limits_V \mu_B(v)/v
$$

Then, as mentioned above, the conditional logical proposition P_1 can be transformed to the F-relation $R(A_1(x), A_2(y))$ by adaptation of the Ply-operator in multi-valued logical system, i.e.

$$
R(A_1(x), A_2(y)) = A \times V \rightarrow U \times B = \int\limits_{U \times V} \left(\mu_A(u) \rightarrow \mu_B(v)\right)/(u, v)
\tag{1.72}
$$

where the values $\mu_A(u) \rightarrow \mu_B(v)$ are depending on the selected logical system.

Assuming $R(A_1(x)) = A$ we can conclude a logical consequence $R(A_2(y))$, then using the CIR for $R(A_1(x))$ and $R(A_1(x), A_2(y))$, then

$$
\begin{aligned}
R(A_2(y)) &= A \circ R(A_1(x), A_2(y)) \\
&= \int\limits_U \mu_A(u)/u \circ \int\limits_{U \times V} \mu_A(u) \rightarrow \mu_B(v))/(u, v) \\
&= \int\limits_V \underset{u \in V}{\vee} [\mu_A(u) \wedge (\mu_A(u) \rightarrow \mu_B(v))]
\end{aligned}
\tag{1.73}
$$

For the criterion I to be satisfied, one of the following equalities must hold true

$$
R(A_2(y)) = B,
$$

$$\bigvee_{u \in V} [\mu_A(u) \wedge (\mu_A(u) \rightarrow \mu_B(v))] = \mu_B(v),$$

or

$$[\mu_A(u) \wedge (\mu_A(u) \rightarrow \mu_B(v))] \leq \mu_B(v) \qquad (1.74)$$

the latter takes place for any $u \in U$ and $v \in V$ or in terms of truth values:

$$T(P \wedge (P \rightarrow Q)) \leq T(Q) \qquad (1.75)$$

The following two conditions are necessary for formalization of F-conditional logical inference rules: the conditional logical inference rules (CIR) must meet the criteria I-IV; the conditional logical inference rules (CIR) satisfy the inequality Eq. 1.75. Now we consider formalization of the fuzzy conditional inference for a different type of conditional propositions. As was shown above, the logical inference for conditional propositions of type P_1 is of the following form:

> Proposition 1 : IF x is A THEN y is B
> Proposition 2 : x is A'
> ----------------------------
> Conclusion : y is B'. $\qquad$ (1.76)

where A, B, and A' are F-concepts represented as F-sets in U, V, and V, respectively, which should satisfy the criteria I, II-1, III, and IV-1.

For this inference if the Proposition 2 is transformed to an unary F-relation in the form $R(A_1(x)) = A'$ and the Proposition 1 is transformed to an F-relation $R(A_1(x), R(A_2(y))$ defined below, then the conclusion $R(A_2(y))$ is derived by using the corresponding F-conditional logical inference rule, i.e.

$$R(A_2(y)) = R(A_1(x)) \circ R(A_1(x)) \qquad (1.77)$$

where $R(A_2(y))$ is equivalent to B' in Eq. 1.76.

1.1.7   Fuzzy Conditional Inference Rule

Theorem 1.2 *If the F-sets A from U and B from V are given in the traditional form:*

$$A = \int_U \mu_A(u)/u, \ B = \int_V \mu_B(v)/v \qquad (1.78)$$

and the relation for the multi-valued logical system ALI1

$$R_1(A_1(x), A_2(y)) = A \times V \underset{ALI1}{\to} U \times B$$

$$= \int_{U \times V} \mu_A(u)/(u, v) \underset{ALI1}{\to} \int_{U \times V} \mu_B(v)/(u, v)$$

$$= \int_{U \times V} \left(\mu_A(u) \underset{ALI1}{\to} \mu_B(v) \right)/(u, v) \tag{1.79}$$

where

$$\mu_A(u) \underset{ALI1}{\to} \mu_B(v) = \left\{ \begin{array}{ll} 1 - \mu_A(u), & \mu_A(u) < \mu_B(v) \\ 1, & \mu_A(u) = \mu_B(v) \\ \mu_B(v), & \mu_A(u) > \mu_B(v) \end{array} \right\}$$

then the criteria I-IV are satisfied.

Below we will consider ALI4 in details.

Consider a continuous function $F(p, q) = p - q$ which defines a distance between p and q where p, q assume values in the unit interval. Notice that $F(p, q) \in [-1, 1]$, where $F(p, q)^{min} = -1$ and $F(p, q)^{max} = 1$. The normalized version of $F(p, q)$ is defined as follow

$$F(p, q)^{norm} = \frac{F(p, q) - F(p, q)^{min}}{F(p, q)^{max} - F(p, q)^{min}} = \frac{F(p, q) + 1}{2} = \frac{p - q + 1}{2} \tag{1.80}$$

It is clear that $F(p, q)^{norm} \in [0, 1]$. This function quantifies a concept of "close-ness" between two values (potentially the ones for the truth values of antecedent and consequent), defined within unit interval, which therefore could play significant role in the formulation of the implication operator in a fuzzy logic.

Definition 1.31 *Implication.* An implication is a continuous function I from $[0, 1] \times [0, 1]$ into $[0, 1]$ such that for $\forall p, p', q, q' r \in [0, 1]$ the following properties (axioms) are satisfied:

11. If $p \le p'$ then $I(p, q) \ge I(p', q)$ (Antitone in first argument),
12. If $q \le q'$ then $I(p, q) \le I(p, q')$ (Monotone in second argument),
13. $I(0, q) = 1$, (Falsity),
14. $I(1, q) \le q$ (Neutrality),
15. $I(p, I(q, r)) = I(q, I(p, r))$ (Exchange),
16. $I(p, q) = I(n(q), n(p))$ (Contra positive symmetry),
      where $n()$ is a negation, which could be defined as $n(q) = T(\neg Q) = 1 - T(Q)$.

Let us define the implication operation

Table 1.4 Fuzzy logic operations

Name	Designation	Value
Tautology	$\overset{\bullet}{P}$	1
Controversy	$\overset{\circ}{P}$	0
Negation	$\neg P$	$1 - P$
Disjunction	$P \vee Q$	$\begin{cases} \dfrac{p+q}{2}, p+q \neq 1, \\ 1, p+q = 1 \end{cases}$
Conjunction	$P \wedge Q$	$\begin{cases} \dfrac{p+q}{2}, p+q \neq 1, \\ 0, p+q = 1 \end{cases}$
Implication	$P \rightarrow Q$	$\begin{cases} \dfrac{1-p+q}{2}, p \neq q, \\ 1, p = q \end{cases}$
Equivalence	$P \leftrightarrow Q$	$\begin{cases} \min((p-q),(q-p)), p \neq q, \\ 1, p = q \end{cases}$
Pierce arrow	$P \downarrow Q$	$\begin{cases} 1 - \dfrac{p+q}{2}, p+q \neq 1, \\ 0, p+q = 1 \end{cases}$
Shaffer stroke	$P \uparrow Q$	$\begin{cases} 1 - \dfrac{p+q}{2}, p+q \neq 1, \\ 1, p+q = 1 \end{cases}$

$$I(p,q) = \left\{ \begin{array}{l} 1 - F(p,q)^{norm}, p > q \\ 1, p \leq q \end{array} \right\} \tag{1.81}$$

where $F(p,q)^{norm}$ is expressed by Eq. 1.80. Before showing that operation $I(p,q)$ satisfies axioms (I1)–(I6), let us show some basic operations encountered in proposed fuzzy logic.

Let us designate the truth values of the *antecedent* P and the *consequent* Q as $T(P) = p$ and $T(P) = q$, respectively. The relevant set of proposed fuzzy logic operators is shown in Table 1.4.

To obtain the truth values of these expressions, we use well known logical properties such as

$p \rightarrow q = \neg p \vee q, p \wedge q = \neg(\neg p \vee \neg q)$ and alike.

In other words, we propose a new many-valued system, characterized by the set of *union* ($\cup$) and *intersection* ($\cap$) operations with relevant *complement*, defined as $T(\neg Q) = 1 - T(Q)$. In addition, the operators $\downarrow$ and $\uparrow$ are expressed as negations of the $\cup$ and $\cup$, respectively. It is well known that the *implication* operation in fuzzy logic supports the foundations of decision-making exploited in numerous schemes of approximate reasoning. Therefore let us prove that the proposed *implication* operation in Eq. 1.81 satisfies axioms (I1)–(I6). For this matter, let us emphasize that we are working with a many-valued system, whose values for our purposes are the elements of the real interval $R = [0, 1]$. For our discussion the set of truth values $V_{11}\{0, 0.1, 0.2, \ldots, 0.9, 1\}$ is sufficient. In further investigations, we use this particular set V_{11}.

Theorem 1.3 *Let a continuous function* I(p, q) *be defined by Eq. 1.73 i.e.*

$$I(p,q) = \left\{ \begin{array}{l} 1 - F(p,q)^{norm}, p > q \\ 1 \qquad\qquad p \leq q \end{array} \right\}, p > q = \left\{ \begin{array}{l} \dfrac{1-p+q}{2}, p > q \\[2mm] 1, p \leq q \end{array} \right. \qquad (1.82)$$

where F(p, q)^{norm} *is defined by Eq. 1.58. Then axioms (I1)–(I6) are satisfied and, therefore Eq. 1.82 is an implication operation.*

It should be mentioned that the proposed fuzzy logic could be characterized by yet some other three features:

$p \wedge 0 \equiv 0$, $p \leq 1$, *whereas* $p \wedge 1 \equiv p$, $p \geq 0$ *and* $\neg\neg p = p$.

As a conclusion, we should admit that all above features confirm that resulting system *can be applied to* V_{11} *for every finite and infinite* n *up to that* $(V_{11}, \neg, \wedge, \vee, \rightarrow)$ *is then* closed *under all its operations.*

Let us investigate Statistical Properties of the Fuzzy Logic. In this section, we discuss some properties of the proposed fuzzy implication operator Eq. 1.82, assuming that the two propositions (*antecedent/consequent*) in a given compound proposition are independent of each other and the truth values of the propositions are uniformly distributed (Li et al. 2010) in the unit interval. In other words, we assume that the propositions P and Q are independent from each other and the truth values $v(P)$ and $v(Q)$ are uniformly distributed across the interval $[0, 1]$. Let $p = v(P)$ and $q = v(Q)$. Then the value of the implication $I = v(p \rightarrow q)$ could be represented as the function $I = I(p, q)$

Because p and q are assumed to be uniformly and independently distributed across $[0, 1]$, the expected value of the implication is

$$E(I) = \iint\limits_{R} I(p, q) dp dq, \qquad (1.83)$$

Its variance is equal to

$$Var(I) = E\left[(I - E(I))^2\right] = \iint\limits_{R} (I(p,q) - E(I))^2 dp dq = E[I^2] - E[I^2] \qquad (1.84)$$

where $R = \{(p, q) : 0 \leq p \leq 1, 0 \leq q \leq 1\}$ From Eq. 1.83 and given Eq. 1.81 as well as the fact that

$$I(p,q) = \left\{ \begin{array}{l} I_1(p,q), p > q, \\ I_2(p,q), p \leq q, \end{array} \right.$$

we have the following

$$E(I_1) = \iint\limits_{\mathfrak{R}} I_1(p,q)dpdp, = \int_0^1\int_0^1 \frac{1-p+q}{2}dpdq = \frac{1}{2}\int_0^1 \left(\int_0^1 (1-p+q)dp\right)dp$$

$$= \frac{1}{2}\left[\int_0^1\left(\left(p - \frac{p^2}{2} + p\right)\Big|_{p=0}^{p=1}\right)dq\right] = \frac{1}{2}\left[\frac{1}{2} + \frac{q^2}{2}\Big|_{q=0}^{q=1}\right] = \frac{1}{2} \qquad (1.85)$$

Whereas $E(I_2) = 1$. Therefore $E(I) = (E(I_1) + E(I_2))/2 = 0.75$
From Eq. 1.84 we have

$$I_1^2(p,q) = \frac{1}{4}(1-p+q)^2 = \frac{1}{4}(1 - 2p + 2q + p^2 - 2pq + q^2)$$

$$E(I_1^2) = \iint\limits_{\mathfrak{R}} I_1^2(p,q)dpdp, = \frac{1}{4}\int_0^1\left(\int_0^1 (1 - 2p + 2q + p^2 - 2pq + q^2)dp\right)dp$$

$$= \frac{1}{4}\int_0^1\left[p - 2\frac{p^2}{2} + \frac{p^3}{3} - 2\frac{p^2}{2}q + 2q + q^2\right]\Big|_{p=0}^{p=1}dq = \frac{1}{4}\int_0^1\left(\frac{1}{3} + q + q^2\right)dq$$

$$= \frac{1}{4}\left[\frac{q}{3} + \frac{q^2}{2} + \frac{q^3}{3}\right]\Big|_{q=0}^{q=1} = \frac{7}{24}$$

Here $E(I_2^2) = 1$ Therefore $E(I^2) = \left(E(I_1^2) + E(I_2^2)\right)/2 = \frac{31}{48}$ From Eqs. 1.84 and 1.85 we have $Var(I) = \frac{1}{12} = 0.0833$

Both values of $E(I)$ and $Var(I)$ demonstrate that the proposed fuzzy implication operator could be considered as one of the fuzziest from the list of the exiting implications (Hu et al. 2010). In addition, it satisfies the set of important Criteria I-IV, which is not the case for the most implication operators mentioned above.

As it was mentioned in Fukami et al. (1980) "in the semantics of natural language there exist a vast array of concepts and humans very often make inferences antecedents and consequences of which contain fuzzy concepts". A formalization of methods for such inferences is one of the most important issues in fuzzy sets theory. For this purpose, let U and V (from now on) be two *universes of discourses* and P and Q are corresponding fuzzy sets:

$$P = \int_U \mu_{\tilde{P}}(u)/u, \quad Q = \int_V \mu_Q(v)/v, \qquad (1.86)$$

Given Eq. 1.86, a *binary relationship* for the fuzzy conditional proposition of the type: "IF x is P THEN y is Q" for proposed fuzzy logic is defined as

$$R(A_1(x), A_2(y)) = P \times V \to U \times B$$

$$\int_{U\times V} \mu_P(u)/(u, v) \to \int_{U\times V} \mu_Q(v)/(u, v) \tag{1.87}$$

$$\int_{U\times V} (\mu_P(u) \to \mu_Q(v))/(u, v)$$

Given Eq. 1.81, expression (1.73) reads as

$$\mu_P(u) \to \mu_Q(v) = \begin{cases} \dfrac{1 - \mu_P(u) + \mu_Q(v)}{2}, \mu_P(u) > \mu_Q(v) \\ 1, \mu_P(u) \le \mu_Q(v) \end{cases} \tag{1.88}$$

It is well known that given a *unary relationship* $R(A_1(x))$ one can obtain the consequence $R(A_2(y))$ by applying a compositional rule of inference (CRI) to $R(A_1(x))$ and $R(A_1(x), A_2(y))$ of type Eq. 1.87:

$$R(A_2(y)) = P \circ R(A_1(x), A_2(y))$$

$$\int_U \mu_P(u)/u \circ \int_{U\times V} \mu_P(u) \to \mu_Q(v)/(u, v) \tag{1.89}$$

$$\int_V \bigcup_{u \in V} [\mu_P(u) \wedge (\mu_P(u) \to \mu_Q(v))]/v$$

In order to have Criterion I satisfied, that is $R(A_2(y)) = Q$ from Eq. 1.89, the equality

$$\int_V \bigcup_{u \in V} [\mu_P(u) \wedge (\mu_P(u) \to \mu_Q(v))] = \mu_Q(v) \tag{1.90}$$

has to be satisfied for any arbitrary v in V. To satisfy Eq. 1.90, it becomes necessary that the inequality

$$\mu_P(u) \wedge (\mu_P(u) \to \mu_Q(v)) \le \mu_Q(v) \tag{1.91}$$

holds for arbitrary $u \in U$ and $v \in V$. Let us define a new method of fuzzy conditional inference of the following type:

$$\begin{aligned} &\text{Ant 1}:\ \text{If } x \text{ is } P \text{ then } y \text{ is } Q \\ &\text{Ant 2}:\ \ x \text{ is } P' \\ &\overline{\hspace{4cm}} \\ &\text{Cons}:\ \text{is } Q' \end{aligned} \tag{1.92}$$

where $P, P \subseteq U$ and $Q, Q' \subseteq V$. Fuzzy conditional inference in the form given by Eq. 1.92 should satisfy Criteria I-IV. It is clear that the inference Eq. 1.92 is defined by the expression (1.89), when $R(A_2(y)) = Q'$.

Theorem 1.4 *If fuzzy sets* $P \subseteq U$ *and* $Q \subseteq V$ *are defined by Eqs. 1.87 and 1.88, respectively and* $R(A_1(x), A_2(y))$ *is expressed as*

$$R(A_1(x), A_2(y)) = P \times V \underset{ALI4}{\rightarrow} U \times Q$$

$$= \int_{U \times V} \mu_P(u)/(u, v) \underset{ALI4}{\rightarrow} \int_{U \times V} \mu_Q(v)/(u, v)$$

$$= \int_{U \times V} \left(\mu_P(u) \underset{ALI4}{\rightarrow} \mu_Q(v) \right)/(u, v)$$

where

$$\mu_P(u) \underset{ALI4}{\rightarrow} \mu_Q(v) = \begin{cases} \dfrac{1 - \mu_P(u) + \mu_Q(v)}{2}, \mu_P(u) > \mu_Q(v) \\ 1, \mu_P(u) \leq \mu_Q(v) \end{cases} \tag{1.93}$$

then Criteria I, II, III and IV-1 (Fukami et al. 1980) are satisfied (Aliev and Tserkovny 2011).

Theorem 1.5 *If fuzzy sets* $P \subseteq U$ *and* $Q \subseteq V$ *are defined by Eqs. 1.87 and 1.88, respectively, and* $R(A_1(x), A_2(y))$ *is defined as*

$$R_1(A_1(x), A_2(y)) = \left(P \times V \underset{ALI4}{\rightarrow} U \times Q \right) \cap \left(\neg P \times V \underset{ALI4}{\rightarrow} U \times \neg Q \right)$$
$$= \int_{U \times V} \left(\mu_P(u) \underset{ALI4}{\rightarrow} \mu_Q(v) \right) \wedge \left((1 - \mu_P(u)) \underset{ALI4}{\rightarrow} (1 - \mu_Q(v)) \right)/(u, v) \tag{1.94}$$

where

$$\left(\mu_P(u) \underset{ALI4}{\rightarrow} \mu_Q(v) \right) \wedge \left((1 - \mu_P(u)) \underset{ALI4}{\rightarrow} (1 - \mu_Q(v)) \right)$$
$$= \begin{cases} \dfrac{1 - \mu_P(u) + \mu_Q(v)}{2}, \mu_P(u) > \mu_Q(v), \\ 1, \mu_P(u) = \mu_Q(v), \\ \dfrac{1 - \mu_P(u) + \mu_Q(v)}{2}, \mu_P(u) < \mu_Q(v), \end{cases}$$

then Criteria I, II, III and IV-2 (Fukami et al. 1980) are satisfied.

Theorems 1.3 and 1.4 show that fuzzy conditional inference rules, defined in Eq. 1.94 could adhere with human intuition to the higher extent as the one defined by Eq. 1.93. The major difference between mentioned methods of inference might be explained by the difference between *Criteria IV-1* and *IV-2*. In particular, a satisfaction of the *Criterion IV-1* means that in case of logical negation of an original antecedent we achieve an ambiguous result of an inference, whereas for the case of the *Criterion IV-2* there is a certainty in a logical inference. Let us to investigate stability and continuity of fuzzy conditional inference in this section. We revisit the fuzzy conditional inference rule (1.92). It will be shown that when the membership function of the observation P is continuous, then the conclusion Q depends continuously on the observation; and when the membership function of the relation R is continuous then the observation Q has a continuous membership function. Let A be a fuzzy number, then for any $\theta \geq 0$ we define $\omega_A(\theta)$ the modulus of continuity of A by

$$\omega_A(\theta) = \max_{|x_1 - x_2| \leq \theta} |\mu_A(x_1) - \mu_A(x_2)| \qquad (1.95)$$

$$D(A,B) = \sup_{\alpha \in [0,1]} d([A]^\alpha, [B]^\alpha) \qquad (1.96)$$

Where d denotes the classical Hausdorff metric expressed in the family of compact subsets of R^2, i.e.

$$d([A^\alpha], [B]^\alpha) = \max\{|a_1(\alpha) - b_1(\alpha)|, |a_2(\alpha) - b_2(\alpha)|\}$$

whereas

$$[A^\alpha] = [a_1(\alpha), a_2(\alpha)], [B]^\alpha = b_1(\alpha), b_2(\alpha).$$

When the fuzzy sets A and B have finite support $\{x_1, \ldots, x_n\}$ then their Hamming distance is defined as

$$H(A,B) = \sum_{i=1}^{n} |\mu_A(x_i) - \mu_B(x_i)|$$

In the sequel we will use the following lemma.

Lemma 1.1 Let $\delta \geq 0$ be a real number and let A, B be fuzzy intervals. If $D(A,B) \leq \delta$, then

$$\sup_{t \in} |\mu_A(t) - \mu_{B'}(t)| \leq \max\{\omega_A(\delta), \omega_B(\delta)\}$$

Consider the fuzzy conditional inference rule with different observations P and P':

Ant 1 : If x is P THEN y is Q

Ant 2 : x is P

Cons : y is Q

Ant 1 : If x is P THEN y is Q

Ant 2 : x is P'

Cons : y is Q'

According to the fuzzy conditional inference rule, the membership functions of the conclusions are computed as

$$\mu_Q(v) = \bigcup_{u \in R} \left[\mu_P(u) \wedge \left(\mu_P(u) \rightarrow \mu_Q(v) \right) \right],$$

$$\mu_{Q'}(v) = \bigcup_{u \in R} \left[\mu_{P'}(u) \wedge \left(\mu_P(u) \rightarrow \mu_Q(v) \right) \right],$$

or

$$\mu_Q(v) = \sup \left[\mu_P(u) \wedge \left(\mu_P(u) \rightarrow \mu_Q(v) \right) \right],$$
$$\mu_{Q'}(v) = \sup \left[\mu_{P'}(u) \wedge \left(\mu_P(u) \rightarrow \mu_Q(v) \right) \right],$$

(1.97)

The following theorem shows the fact that when the observations are closed to each other in the metric $D(.)$ of Eq. 1.96 type, then there can be only a small deviation in the membership functions of the conclusions.

Theorem 1.6 *(Stability theorem) (Aliev and Tserkovny 2011). Let $\delta \geq 0$ and let P, P′ be fuzzy intervals and an implication operation in the fuzzy conditional inference rule (1.97) is of type Eq. 1.88. If $D(P, P') \leq \delta$, then*

$$\sup_{v \in R} \left| \mu_P(v) - \mu_{P'}(v) \right| \leq \max\{\omega_P(\delta), \omega_{P'}(\delta)\}$$

Theorem 1.7 *(Continuity theorem) (Aliev and Tserkovny 2011). Let binary relationship $R(u, v) = = \mu_p(u) \rightarrow_{ALI4} \mu_Q(v)$ be continuous. Then Q is continuous and $\omega_Q(\delta) \leq \omega_R(\delta)$ for each $\delta \geq 0$.*

1.2 Type-2 Fuzzy Sets

In many real-world pattern recognition, identification and modeling, control, decision making, forecasting of time-series and other applications, it may be impossible in most cases to obtain perfect information. An emerging approach is to employ fuzzy type 2 sets which possess more expressive power to create models adequately describing such uncertainty. The three-dimensional membership functions of type-

2 fuzzy sets provide additional degrees of freedom that make it possible to directly and more effectively model uncertainties.

Definition 1.32 *Type-2 fuzzy set.* (Karnik et al. 1999; Karnik and Mendel 2001; Mendel and John 2002; Mendel 2007; Castillo and Melin 2008a) Type-2 fuzzy set in the universe of discourse X can be represented by a type-2 membership function $\mu\widetilde{A}$ shown as follows:

$$\widetilde{A} = \left\{ ((x,u), \mu_{\widetilde{A}}(x,u)) \,|\, \forall x \in X, \forall u \in J_x \subseteq [0,1] \right\} \qquad (1.98)$$

where $0 \leq \mu\widetilde{A}(x,u) \leq 1$.

A type-2 fuzzy set $\widetilde{A}$ can also be represented as follows:

$$\widetilde{A} = \int_{x \in X} \int_{u \in J_x} \mu_{\widetilde{A}}(x,u)/(x,u) = \int_{x \in X} \left[\int_{u \in J_x} \mu_{\widetilde{A}}(x,u)/u \right] /x \qquad (1.99)$$

where x is the primary variable, $J_x \subseteq [0,1]$ is the primary membership of x, u is the secondary variable, and $f(x) = \int_{u \in J_x} \mu_{\widetilde{A}}(x,u)/u$ is the secondary membership function at x. $\int\int$ denotes the union over all admissible x and u.

If X is a discrete set with elements $x_1, \ldots, x_n$ then type-2 fuzzy set $\widetilde{A}$ can be represented as follows:

$$\widetilde{A} = \sum_{x \in X} \left[\sum_{u \in J_x} f_x(u)/u \right] /x = \sum_{i=1}^{n} \sum_{u \in J_x} [f_x(u)/u]/x_i \qquad (1.100)$$

Some examples of type-2 membership functions are given in Fig. 1.8. The shaded area referred to as footprint of uncertainty (FOU) and implies that there is a distribution that indicates the third dimension in type-2 fuzzy sets.

Definition 1.33 *Footprint of uncertainty.* To indicate second order uncertainty one can use the concept of the footprint of uncertainty. The footprint of uncertainty (FOU) of a type-2 fuzzy set $\widetilde{A}$ is a region with boundaries covering all the primary membership points of elements x, and is defined as follows: (Mendel 2001, 2003, 2007; Castillo 2011; Castillo and Melin 2008a, b; Jaffal and Tao 2011).

$$\text{FOU}\left(\widetilde{A}\right) = \bigcup_{x \in X} J_x \qquad (1.101)$$

Example Fuzzy set of type-2 is shown in Fig. 1.9. If $x = x_1$ four numbers $\alpha_1, \alpha_2, \alpha_3, \alpha_4$ are produced, by which the ordinary fuzzy set defined with trapezoidal membership function assigned to x_1 is determined.

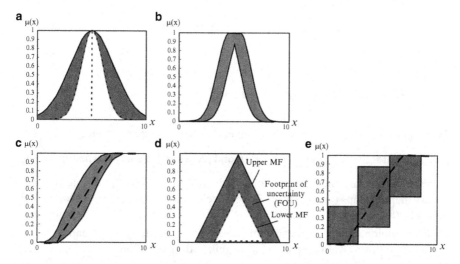

Fig. 1.8 Examples of MFs and FOUs: (**a**) Gaussian MF with uncertain standard deviation; (**b**) Gaussian MF with uncertain mean; (**c**) Sigmoidal MF with inflection uncertainties; (**d**) Triangle type-2 MF; (**e**) Granulated sigmoidal MF with granulation uncertainties

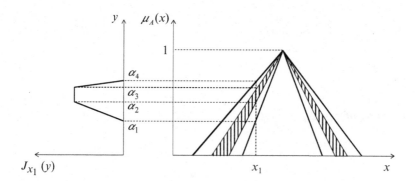

Fig. 1.9 Fuzzy set of type-2

Definition 1.34 *Embedded fuzzy sets.* The embedded type-2 fuzzy set of type-2 fuzzy set $\tilde{A}$ is defined as follows (Mendel and John 2002):

$$\tilde{A}_O = \int\limits_{x \in X} [f_x(\theta)/\theta]/x, \qquad (1.102)$$

where θ is the element, which can be chosen from each interval J_x. For discrete case $\tilde{A}_O$ is defined as follows:

$$\widetilde{A}_O = \sum_{i=1}^{R} \left[f_{x_i}(\theta_i)/\theta_i \right]/x_i. \tag{1.103}$$

1.2.1 Operations on Type-2 Fuzzy Sets

Two type-2 fuzzy sets $\widetilde{A}$ and $\widetilde{B}$, in a universe X with membership functions $\mu_{\widetilde{A}}(x)$ and $\mu_{\widetilde{B}}(x)$ are given. $\mu_{\widetilde{A}}(x) = \int_u f_x(u)/u$ and $\mu_{\widetilde{B}}(x) = \int_w g_x(w)/w$ where $u, w \subseteq J_x$ indicate the primary memberships of x and $f_x(u), g_x(w) \in [0,1]$ indicate the secondary memberships (grades) of x. Using Zadeh's Extension Principle (Karnik and Mendel 2001; Aliev et al. 1991a; Aliev and Tserkovny 1988), the membership grades for union, intersection and complement of type-2 fuzzy sets $\widetilde{A}$ and $\widetilde{B}$ can be defined as follows (Karnik et al. 1998; Karnik and Mendel 2001):

Union of type-2 fuzzy sets:

$$\widetilde{A} \cup \widetilde{B} \Leftrightarrow \mu_{\widetilde{A} \cup \widetilde{B}} = \mu_{\widetilde{A}}(x) \sqcup \mu_{\widetilde{B}}(x) = \int_u \int_w (f_x(u) * g_x(w))/(u \vee w), \tag{1.104}$$

Intersection of type-2 fuzzy sets:

$$\widetilde{A} \cap \widetilde{B} \Leftrightarrow \mu_{\widetilde{A} \cap \widetilde{B}} = \mu_{\widetilde{A}}(x) \sqcap \mu_{\widetilde{B}}(x) = \int_u \int_w (f_x(u) * g_x(w))/(u * w), \tag{1.105}$$

Complement of type-2 fuzzy set:

$$\overline{\widetilde{A}} \Leftrightarrow \mu\overline{\widetilde{A}} = \neg\mu_{\widetilde{A}}(x) = \int_u f_x(u)/(1 - u). \tag{1.106}$$

Here $\sqcap$ and $\sqcup$ are intersection/meet and union/join operations on two membership function type-2 fuzzy sets, $*$ indicates the chosen T-norm.

T-norm and T-conorm

A T-norm can be extended to be a conjunction in type-2 logic and an intersection in type-2 fuzzy set theory, such as a Minimum T-norm and a Product T-norm (Mendel 2008).

A T-conorm of operation can be used to stand for a disjunction in type-2 fuzzy logic and a union in type-2 fuzzy set theory, such as maximum T-conorm.

1.2.2 Type-2 Fuzzy Relations

Let $X_1, X_2, \ldots, X_n$ be n universes. A type-2 fuzzy relation in $X_1 \times X_2 \times \ldots \times X_n$ is type-2 fuzzy subset of the Cartesian product space.

Let X and Y based on the Cartesian product $X*Y$ are given. The binary type-2 fuzzy relation $\widetilde{R}$ between X and Y can be defined as follows:

$$\widetilde{R}(X, Y) = \int_{X*Y} \mu_{\widetilde{R}}(x, y)/(x, y), \tag{1.107}$$

where $x \in X$, $y \in Y$.

The membership function of (x,y) is given as follows:

$$\mu_{\widetilde{R}}(x, y) = \int_{v \in J_{x,y}^v} r_{x,y}(v)/v,$$

where $r_{x,y}(v)$ is the secondary membership and $J_{x,y}^v \subset [0, 1]$.

Let $\widetilde{R}(X, Y)$ and $\widetilde{S}(X, Y)$ are two type-2 fuzzy relation on the same product space $X \times Y$. Their union and intersection can be defined as follows:

$$\mu_{\widetilde{R} \cup \widetilde{S}}(x, y) = \mu_{\widetilde{R}}(x, y) \sqcup \mu_{\widetilde{S}}(x, y) \tag{1.108}$$

$$\mu_{\widetilde{R} \cap \widetilde{S}}(x, y) = \mu_{\widetilde{R}}(x, y) \sqcap \mu_{\widetilde{S}}(x, y) \tag{1.109}$$

Type Reduction

To defuzzify type-2 fuzzy sets one can use type reduction procedure. By this procedure, we can transform a type-2 fuzzy set into a type-1 fuzzy set. The centroid of a type-2 set, whose domain is discretized into points, can be defined as follows:

$$C_{\widetilde{A}} = \int_{\theta_1} \ldots \int_{\theta_N} \left[\mu_{D_1}(\theta_1) * \ldots * \mu_{D_N}(\theta_N)\right] / \frac{\sum\limits_{i=1}^{N} x_i \theta_i}{\sum\limits_{i=1}^{N} \theta_i} \tag{1.110}$$

where $D_i = \mu_{\widetilde{A}}(x_i)$, $\theta_i \in D_i$.

Composition of Type-2 Fuzzy Relations

Sur-star composition of two type-2 fuzzy relations $\tilde{R}$ and $\tilde{S}$ is determined as

$$\mu_{\underset{R \circ S}{\sim}}(u,w) = \bigsqcup_{v \in V} \left[\mu_{\tilde{R}}(u,v) \sqcap \mu_{\tilde{R}}(v,w) \right].$$

1.2.3 Type-2 Fuzzy Number

Let $\tilde{A}$ be a type-2 fuzzy set defined in the universe of discourse R. If $\tilde{A}$ is normal, $\tilde{A}$ is a convex set, and the support of $\tilde{A}$ is closed and bounded, then $\tilde{A}$ is a type-2 fuzzy number (Dinagar and Anbalagan 2011).

We concentrate on triangular type-2 fuzzy numbers.

A type-2 triangular fuzzy number $\tilde{A} = (A_1, A_2, A_3)$ on R is given by

$$\tilde{A} = \left\{ \left(x, \left(\mu_A^1(x), \mu_A^2(x), \mu_A^3(x)\right)\right); x \in R \right\} \text{ and}$$
$$\mu_A^1(x) \le \mu_A^2(x) \le \mu_A^3(x),$$

for all $x \in R$, $A_1 = (A_1^L, A_1^N, A_1^U)$, $A_2 = (A_2^L, A_2^N, A_2^U)$ and $A_1 = (A_3^L, A_3^N, A_3^U)$.

Let's consider arithmetic operations on type-2 triangular fuzzy numbers (Dinagar and Latha 2012).

Assume two type-2 triangular fuzzy numbers

$$\tilde{a} = (a_1, a_2, a_3) = \left((a_1^L, a_1^N, a_1^U), (a_2^L, a_2^N, a_2^U), (a_3^L, a_3^N, a_3^U) \right) \text{ and}$$
$$\tilde{b} = (b_1, b_2, b_3) = \left((b_1^L, b_1^N, b_1^U), (b_2^L, b_2^N, b_2^U), (b_3^L, b_3^N, b_3^U) \right)$$

are given.

Addition of type-2 triangular fuzzy numbers $\tilde{a}$ and $\tilde{b}$ is determined as

$$\tilde{a} + \tilde{b} = \left((a_1^L + b_1^L, a_1^N + b_1^N, a_1^U + b_1^U), (a_2^L + b_2^L, a_2^N + b_2^N, a_2^U + b_2^U), \right.$$
$$\left. (a_3^L + b_3^L, a_3^N + b_3^N, a_3^U + b_3^U) \right).$$

Subtraction of type-2 triangular fuzzy numbers $\tilde{a}$ and $\tilde{b}$ can be defined as

$$\tilde{a} - \tilde{b} = \left((a_1^L - b_3^U, a_1^N - b_3^N, a_1^U - b_3^L), (a_2^L - b_2^L, a_2^N - b_2^N, a_2^U - b_2^U), \right.$$
$$\left. (a_3^L - b_1^U, a_3^N - b_1^N, a_3^U - b_1^L) \right).$$

Scalar multiplication of type-2 triangular fuzzy number $\tilde{a}$ is defined as
If $k \ge 0$ and $k \in R$ then

$$k\widetilde{a} = \left(\left(ka_1^L, ka_1^N, ka_1^U \right), \left(ka_2^L, ka_2^N, ka_2^U \right), \left(ka_3^L, ka_3^N, ka_3^U \right) \right)$$

and if $k < 0$ and $k \in R$ then

$$k\widetilde{a} = \left(\left(ka_3^U, ka_3^N, ka_3^L \right), \left(ka_2^U, ka_2^N, ka_2^L \right), \left(ka_1^U, ka_1^N, ka_1^L \right) \right).$$

Let's define $\sigma b = b_1^L + b_1^N + b_1^U + b_2^L + b_2^N + b_2^U + b_3^L + b_3^N + b_3^U$.

Multiplication of type-2 triangular fuzzy numbers $\widetilde{a}$ and $\widetilde{b}$ is determined as follows.

If $\sigma b \geq 0$, then

$$\widetilde{a} \times \widetilde{b} = \left(\left(\frac{a_1^L \sigma b}{9}, \frac{a_1^N \sigma b}{9}, \frac{a_1^U \sigma b}{9} \right), \left(\frac{a_2^L \sigma b}{9}, \frac{a_2^N \sigma b}{9}, \frac{a_2^U \sigma b}{9} \right), \left(\frac{a_3^L \sigma b}{9}, \frac{a_3^N \sigma b}{9}, \frac{a_3^U \sigma b}{9} \right) \right).$$

If $\sigma b < 0$, then

$$\widetilde{a} \times \widetilde{b} = \left(\left(\frac{a_3^U \sigma b}{9}, \frac{a_3^N \sigma b}{9}, \frac{a_3^L \sigma b}{9} \right), \left(\frac{a_2^U \sigma b}{9}, \frac{a_2^N \sigma b}{9}, \frac{a_2^L \sigma b}{9} \right), \left(\frac{a_1^U \sigma b}{9}, \frac{a_1^N \sigma b}{9}, \frac{a_1^L \sigma b}{9} \right) \right).$$

Division of type-2 triangular fuzzy numbers $\widetilde{a}$ and $\widetilde{b}$ is calculated as follows.

Whenever $\sigma b \neq 0$

If $\sigma b > 0$, then

$$\frac{\widetilde{a}}{\widetilde{b}} = \left(\left(\frac{9a_1^L}{\sigma b}, \frac{9a_1^N}{\sigma b}, \frac{9a_1^U}{\sigma b} \right), \left(\frac{9a_2^L}{\sigma b}, \frac{9a_2^N}{\sigma b}, \frac{9a_2^U}{\sigma b} \right), \left(\frac{9a_3^L}{\sigma b}, \frac{9a_3^N}{\sigma b}, \frac{9a_3^U}{\sigma b} \right) \right).$$

If $\sigma b < 0$, then

$$\frac{\widetilde{a}}{\widetilde{b}} = \left(\left(\frac{9a_3^U}{\sigma b}, \frac{9a_3^N}{\sigma b}, \frac{9a_3^L}{\sigma b} \right), \left(\frac{9a_2^U}{\sigma b}, \frac{9a_2^N}{\sigma b}, \frac{9a_2^L}{\sigma b} \right), \left(\frac{9a_1^U}{\sigma b}, \frac{9a_1^N}{\sigma b}, \frac{9a_1^L}{\sigma b} \right) \right).$$

Inverse of type-2 triangular fuzzy number $\widetilde{a}$ is determined as $\widetilde{a}^{-1} = \frac{1}{a}$. $\widetilde{a} \neq 0$

A type-2 triangular fuzzy number matrix (T2TFM) of order $m \times n$ defined as $A = \left(\widetilde{a}_{ij} \right)_{m \times n}$ where the ij-th element $\widetilde{a}_{ij}$ of A is the type-2 triangular fuzzy number. Let briefly consider operations on type-2 triangular fuzzy matrixes.

$A = \left(\widetilde{a}_{ij} \right)$ and $B = \left(\widetilde{b}_{ij} \right)$ are two fuzzy matrixes of same order. Operations on then are as following:

Addition

$$A + B = \left(\widetilde{a}_{ij} + \widetilde{b}_{ij} \right).$$

Subtraction

$$A - B = \left(\tilde{a}_{ij} - \tilde{b}_{ij} \right).$$

Multiplication

For $A = \left(\tilde{a}_{ij} \right)_{m \times n}$ and $B = \left(\tilde{b}_{ij} \right)_{n \times k}$,

$$AB = \left(\tilde{c}_{ij} \right)_{m \times k} \text{ where } \tilde{c}_{ij} = \sum_{p=1}^{n} \tilde{a}_{ip} \cdot \tilde{b}_{pj}, \quad i = 1, 2, \ldots, m \text{ and } j = 1, 2, \ldots, k.$$

Transporation

$$A^T = \left(\tilde{a}_{ji} \right).$$

1.2.4   Interval Type-2 Fuzzy Sets

The computational complexity of dealing with general type-2 fuzzy sets is rather high. To simplify computation, interval type-2 fuzzy sets are commonly used.

As it was mentioned above a general type-2 fuzzy set is described as follows.

$$\tilde{A} = \int_{x \in X} \int_{u \in J_x} \mu_{\tilde{A}} (x, u) / (x, u), \quad J_x \subseteq [0, 1],$$

where $\int\int$ denotes union over all admissible x and u. For discrete universes of discourse, $\int$ is replaced by $\sum$.

Definition 1.35   Interval Type-2 fuzzy set. When all $\mu \tilde{\tilde{A}} (x, u) = 1$ then $\tilde{A}$ is an interval type-2 fuzzy set (IT2 FS).

Consequently IT2 FS can be expressed as

$$\tilde{A} = \int_{x \in X} \int_{u \in J_x} 1 / (x, u), \quad J_x \subseteq [0, 1]. \tag{1.111}$$

Graphical representation of Interval type-2 Fuzzy set is given in Fig. 1.10 (Aliev and Aliev 2001).

Example

$$\tilde{A} = \mu_{\tilde{A}} \left(x_1^{\alpha_1}, x_1^{\alpha_2} \right) / x_1 + \ldots + \mu_{\tilde{A}} \left(x_n^{\alpha_1}, x_n^{\alpha_2} \right) / x_n$$

The membership function of interval-valued fuzzy set $\tilde{A}$ (Aliev and Aliev 2001) is given in Fig. 1.11.

Fig. 1.10 Interval type-2
Fuzzy set

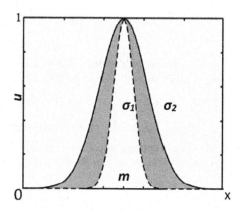

Fig. 1.11 Interval type-2
Fuzzy set

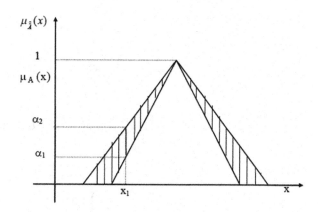

At each value of x, say $x = x'$, the 2-D plane whose axes are u and $\mu_{\tilde{A}}(x', u)$ is
called a vertical slice of $\mu_{\tilde{A}}(x, u)$.

Based on the concept of secondary sets, we can reinterpret an IT2 FS as the
union of all secondary sets, i.e. using Eqs. 1.98 and 1.99, we can re-express $\tilde{A}$ in a
vertical-slice manner, as (Mendel et al. 2006)

$$\tilde{A} = \left\{ (x, \mu_{\tilde{A}}(x)) \,\middle|\, \forall X \in X \right\}$$

or, alternatively, as

$$\tilde{A} = \int_{x \in X} \mu_{\tilde{A}}(x)/x = \int_{x \in X} \left[\int_{u \in J_x} 1/u \right]/x, \quad J_x \subseteq [0, 1].$$

IT2 FS footprint of uncertainty can be described as

$$\mathrm{FOU}\left(\widetilde{A}\right) = \underset{x \in X}{\cup} J_x. \tag{1.112}$$

In accordance with Mendel-John representation theorem for an IT2 FS $\widetilde{A}$ is the union of all of its embedded IT2 FSs.

Assume that $X = \{x_1, x_2, \ldots, x_n\}$ is discrete or discretized primary variable, $U_i^j \in \left\{\underline{\mu}_{\widetilde{A}}(x_i), \ldots, \overline{\mu}_{\widetilde{A}}(x_i)\right\}$ is sampled secondary variables. In accordance with Mendel-John representation Interval type-2 fuzzy sets (IT2FS) can be expressed as follows:

$$\widetilde{A} = 1/FOU\left(\widetilde{A}\right) = 1/U_{j=1}^{n_A} A_e^j \tag{1.113}$$

where $A_e^j = \displaystyle\sum_{i-1}^{n} U_i^j / x_i$

$$\widetilde{A} = \sum_{j=1}^{n_A} \widetilde{A}_e^j$$

$$\widetilde{A}_e^j = \sum_{i=1}^{N} \left[1/u_i^j\right]/x_i, \ \ u_i^j \in J_{x_i} \subseteq U = [0, 1]$$

and

$$n_A = \prod_{i=1}^{N} M_i$$

in which M_i denotes the discretization levels of secondary variable u_i^j at each of the $N \ x_i$.

Operations on IT2 FS. In Mendel et al. (2006) theorem on union, intersection, and complement has been proofed.

Let consider the set-theoretic operations of union, intersection, and complement and the arithmetic operations of addition and multiplication on IT2FS (Mendel et al. 2006).

IT2FS $\widetilde{A}$ and $\widetilde{B}$ are given:

$$\widetilde{A} = \int_X \left[\int_{J_x^u} 1/u\right]/x$$

$$\widetilde{B} = \int_X \left[\int_{J_x^w} 1/w\right]/x$$

Union/join

$$\mu_{\widetilde{A \cup B}}(x) = \left[\underline{\mu_{\widetilde{A}}}(x) \vee \underline{\mu_{\widetilde{B}}}(x), \overline{\mu_{\widetilde{A}}}(x) \vee \overline{\mu_{\widetilde{B}}}(x) \right] \quad \forall x \in X \tag{1.114}$$

Intersection/meet

$$\mu_{\widetilde{A \cap B}}(x) = \left[\underline{\mu_{\widetilde{A}}}(x) \wedge \underline{\mu_{\widetilde{B}}}(x), \overline{\mu_{\widetilde{A}}}(x) \wedge \overline{\mu_{\widetilde{B}}}(x) \right] \quad \forall x \in X \tag{1.115}$$

Complement/negation

$$\mu_{\widetilde{A}}(x) = \left[1 - \overline{\mu_{\widetilde{A}}}(x), 1 - \underline{\mu_{\widetilde{A}}}(x) \right] \quad \forall x \in X \tag{1.116}$$

Addition

$$F + G = \left[\underline{\mu_F} + \underline{\mu_G}, \overline{\mu}_F + \overline{\mu}_G \right] \tag{1.117}$$

Multiplication

$$F \times G = \left[\underline{\mu_F} T \underline{\mu_G}, \overline{\mu}_F T \overline{\mu}_G \right] \tag{1.118}$$

Here $\overline{\mu_{\widetilde{A}}}(x)$ and $\underline{\mu_{\widetilde{A}}}(x)$ are the upper and lower membership function respectively and can be determined as

$$\overline{\mu_{\widetilde{A}}}(x) \equiv \overline{\mathrm{FOU}\left(\widetilde{A}\right)} \quad \forall x \in X$$
$$\underline{\mu_{\widetilde{A}}}(x) \equiv \underline{\mathrm{FOU}\left(\widetilde{A}\right)} \quad \forall x \in X.$$

Arithmetic Operations on IT2 Fuzzy Numbers

Type-2 fuzzy number (T2FN) is very frequently used in type-2 fuzzy decision making and control. We concentrate on triangular type-2 fuzzy number (TIT2FN) The TIT2FN can be determined as follows:

$$\widetilde{a}_i = \left(a_i^l, a_i^u \right) = \left(\left[a_{i1}^l, a_{i2}^l \right], \left[a_{i1}^u, a_{i2}^u \right]; s_i^l, s_i^u \right),$$

where a_i^l and a_i^u are type-2 fuzzy sets, $a_{i1}^l, a_{i2}^l, a_{i1}^u, a_{i2}^u$ are the reference points of the interval type-2 fuzzy set $\widetilde{a}_i, s_i^l$ is the upper membership function and s_i^u is the lower membership function, $s_i^l \in [0,1]$ and $s_i^u \in [0,1]$ and $1 \leq i \leq n$.

Let consider arithmetic operations on TIT2FN (Zamri et al. 2013).

Addition

Two TIT2FN are given

$$\widetilde{a}_1 = \left(a_1^l, a_1^u\right) = \left(\left[a_{11}^l, a_{12}^l\right], \left[a_{11}^u, a_{12}^u\right]; s_1^l, s_1^u\right)$$
$$\widetilde{b}_2 = \left(b_2^l, b_2^u\right) = \left(\left[b_{21}^l, b_{22}^l\right], \left[b_{21}^u, b_{22}^u\right]; s_2^l, s_2^u\right)$$

The addition of $\widetilde{a}_1$ and $\widetilde{b}_2$ is defined as:

$$\widetilde{a}_1 + \widetilde{b}_2 = \left(a_1^l, a_1^u\right) + \left(b_2^l, b_2^u\right) = \begin{bmatrix} \left(a_{11}^l + b_{21}^l, a_{12}^l + b_{22}^l; \min\left(s_1^l, s_2^l\right)\right), \\ \left(a_{11}^u + b_{21}^u, a_{12}^u + b_{22}^u; \min\left(s_1^u, s_2^u\right)\right) \end{bmatrix}$$

Subtraction

The subtraction of $\widetilde{a}_1$ and $\widetilde{b}_2$ is defined as:

$$\widetilde{a}_1 - \widetilde{b}_2 = \left(a_1^l, a_1^u\right) - \left(b_2^l, b_2^u\right) = \begin{bmatrix} \left(a_{11}^l - b_{21}^l, a_{12}^l - b_{22}^l; \min\left(s_1^l, s_2^l\right)\right), \\ \left(a_{11}^u - b_{21}^{ul}, a_{12}^u - b_{22}^u; \min\left(s_1^u, s_2^u\right)\right) \end{bmatrix}$$

Multiplication

The multiplication of $\widetilde{a}_1$ and $\widetilde{b}_2$ is defined as:

$$\widetilde{a}_1 \times \widetilde{b}_2 = \left(a_1^l, a_1^u\right) \times \left(b_2^l, b_2^u\right) = \begin{bmatrix} \left(a_{11}^l \times b_{21}^l, a_{12}^l \times b_{22}^l; \min\left(s_1^l, s_2^l\right)\right), \\ \left(a_{11}^u \times b_{21}^{ul}, a_{12}^u \times b_{22}^u; \min\left(s_1^u, s_2^u\right)\right) \end{bmatrix}$$

Division

The division of $\widetilde{a}_1$ and $\widetilde{b}_2$ is defined as:

$$\widetilde{a}_1 \div \widetilde{b}_2 = \left(a_1^l, a_1^u\right) \div \left(b_2^l, b_2^u\right) = \begin{bmatrix} \left(a_{11}^l \div b_{21}^l, a_{12}^l \div b_{22}^l; \min\left(s_1^l, s_2^l\right)\right), \\ \left(a_{11}^u \div b_{21}^{ul}, a_{12}^u \div b_{22}^u; \min\left(s_1^u, s_2^u\right)\right) \end{bmatrix}$$

1.2.5   Type-2 Fuzzy Logic System (FLS)

Ordinary fuzzy logic systems (i.e. type-1 fuzzy logic systems) have been successfully used widely in various applications. To design ordinary fuzzy logic systems, knowledge of human experts and data are needed for construction of fuzzy rules and membership functions based on available linguistic or numeric information (Aliev and Aliev 2001). However, in many cases available information or data are associated with various types of uncertainty which should be taken into account. The uncertainty can be captured by using higher order fuzzy sets. In this regards, fuzzy type-2 sets can represent and handle uncertain information more effectively than fuzzy type-1 sets and contribute to the robustness and stability of the inference.

Figure 1.12 describes the structure of a type-2 fuzzy logic system (FLS).

A type-2 FLS in general consists of four components: fuzzifier (mapping input into type-2 fuzzy sets), fuzzy type-2 rule base, inference engine, and output

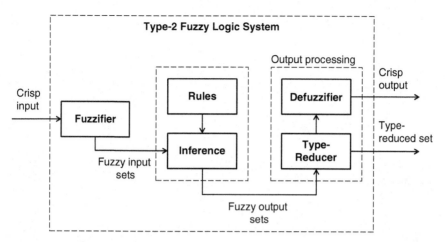

Fig. 1.12 Structure of a type-2 FLS

processor. Fuzzy rule base is expressed in If-Then statements, the antecedents and consequents of which are type-2 fuzzy sets. The j-th fuzzy rule that has n antecedents can be expressed as:

$$R^j : \text{ If } x_1 \text{ is } \tilde{A}_1^j \text{ and } x_2 \text{ is } \tilde{A}_2^j \text{ and } \ldots x_n \text{ is } \tilde{A}_n^j \text{ Then } y \text{ is } \tilde{B}^j, \quad j = \overline{1, m} \quad (1.119)$$

where $x_i \; (i = \overline{1, n})$ and y are input and output variables, respectively; $A_i^j (i = \overline{1, n})$ and $\tilde{B}^j$ are antecedent and consequent type-2 fuzzy sets, respectively.

Inference engine of fuzzy type-2 FLS for inference uses the extended sup-star composition. Input type-2 fuzzy sets in antecedent part of a rule are connected by meet operation whereas output type-2 fuzzy sets of a rule are combined using joint operation (Karnik et al. 1999; Liang and Mendel 2000).

Output processor of type-2 FLS includes two-phase defuzzification. Type reducer converts type-2 fuzzy output sets to type-1 output fuzzy sets. An ordinary defuzzifier do further conversion of type-2 fuzzy sets into crisp values using the same methods used in a type-1 FLS.

Fuzzification

In case of type-1 fuzzy logic systems the inputs can be type-0 (crisp) or a type-1 FS. For the first case

$$\mu_{A_i}(x_i) = \begin{cases} 1, & \text{if } x_i = x_j' \\ 0, & \text{if } x_i \neq x', \end{cases} \quad (1.120)$$

where $x_i \in X_i, \; i = 1, \ldots, n$ is input variables.

For the second case

$$\mu_{A_i}(x_i) = \begin{cases} 1/1, & \text{if } x_i = x_j' \\ 1/0, & \text{if } x_i \neq x' \end{cases} \tag{1.121}$$

On the base of Eqs. 1.20 and 1.21 one finds the only point x' which is the i-th input variable with type-2 fuzzy sets:

$$\tilde{A}_i = (1/1)/x_i', \quad i = 1, \ldots n.$$

Inference in a Type-2 FLS

Assume that a type-2 FLS has n inputs, $x_i \in X_1$, $x_2 \in X_2, \ldots, x_n \in X_n$ and one output $y \in Y$. Let us suppose that it has m rules where the i-th rule has the form (1.119).

The rule (1.119) represents a type-2 fuzzy relation between the input space $X_1 \times X_2 \times \cdots \times X_n$ and the output space Y of the FLS.

The rule (1.119) can be translated into the following form:

$$\tilde{R}^j = \tilde{A}^j \rightarrow \tilde{B}^j, \tag{1.122}$$

where the membership function of $\tilde{A}$ is defined as follows:

$$\mu_{\tilde{A}}^j(x) = T_{i=1}\mu_{\tilde{A}_i}^j(x_i). \tag{1.123}$$

In design of inference engine of type-2 FLS we need to determine membership function $\mu_{\tilde{A}^j \rightarrow \tilde{B}^j}(X, y)$. The j-th rule in the rule base of type-2 FLS can be expressed as follows (Karnik and Mendel 1998a, b; Karnik et al. 1999):

$$\tilde{R}^j(x, y) = \int_{X*Y} \mu_{\tilde{R}^j}(X, y)/(X, y), \tag{1.124}$$

So

$$\mu_{\tilde{A}^j \rightarrow \tilde{B}^j}(X, y) = \mu_{\tilde{R}^{(j)}}(X, y),$$

where the membership function can be expressed by the extended T-norm operation as follows:

$$\mu_{\tilde{A}^j \rightarrow \tilde{B}^j}(X, y) = \mu_{\tilde{A}^j}(X)T\mu_{\tilde{B}^j}(y).$$

The output of inference of type-2 FLS is a type-2 fuzzy set $\tilde{B}^{\prime j}$ can be determined as follows:

$$\mu_{\widetilde{A}^j \to \widetilde{B}^j}(X, y) = \mu_{\widetilde{A}^j}(X) T \mu_{\widetilde{B}^j}(y). \tag{1.125}$$

Here $\widetilde{A}'$ is input type-2 fuzzy set.

In terms of membership function determination of $\widetilde{B}'^j$,

$$\widetilde{B}'^j = \widetilde{A}' \circ \widetilde{R}^j = \widetilde{A}' \circ \left(\widetilde{A}^j \to \widetilde{B}^j \right)$$

can be rewritten as follows:

$$\mu_{\widetilde{B}'^j}(y) = \mu_{\widetilde{A}' \circ \widetilde{R}^j}(y) = \widetilde{S}_{x \in X} \left(\mu_{\widetilde{A}^j}(X) T \mu_{\widetilde{B}^j}(y) \right) = \widetilde{S}_{x \in X} \left(\mu_{\widetilde{A}^j}(X) T \mu_{\widetilde{A}^j \to \widetilde{B}^j}(X, y) \right).$$

And in case of the fuzzification in type-2 fuzzy systems, the membership function can be expressed as follows:

$$\mu_{\widetilde{B}'^j}(y) = \mu_{\widetilde{A}^j \to \widetilde{B}^j}(, \overline{X}, y).$$

Considering Eq. 1.23 and $\mu_{\widetilde{A}^j \to \widetilde{B}^j}(X, y) = \mu_{\widetilde{R}^{(j)}}(X, y)$, the membership function also can take the following form:

$$\mu_{\widetilde{B}'^j}(y) = \mu_{\widetilde{A}^j}(\overline{x}) T \mu_{\widetilde{B}^j}(y) = \left(\widetilde{T}_{i=1}^n \mu_{\widetilde{A}_i^j}(\overline{x}) \right) T \mu_{\widetilde{B}^j}(y).$$

Let τ_j represents the firing strength of j-th rule (1.119).

$$\tau_j = \widetilde{T}_{i=1}^n \mu_{\widetilde{A}_i^j}(\overline{x}).$$

Thus, $\mu_{\widetilde{B}'^j}(y)$ can be defined as

$$\mu_{\widetilde{B}'^j}(y) = \tau_j {}_{*T} \mu_{\widetilde{B}^j}(y).$$

Using the aggregation procedure as extended T-conorm for all rules we get:

$$\mu \widetilde{B}(y) = \widetilde{S}_{j=1}^N \mu_{\widetilde{B}^j}(y)$$

Let's consider reasoning in IT2FLS with triangular Interval type-2 fuzzy numbers. In Fig. 1.13 it is shown graphical representation of calculation of rule output. Assume that $x_1 = x_1'$, the vertical line at x_1' intersects FOU $\left(\widetilde{A}_1 \right)$ everywhere in the interval $\left[\underline{\mu}_{\widetilde{A}_1}(x_1'), \overline{\mu}_{\widetilde{A}_1}(x_1') \right]$; and, when $x_2 = x_2'$, the vertical line at x_2' intersects FOU $\left(\widetilde{A}_2 \right)$ everywhere in the interval $\left[\underline{\mu}_{\widetilde{A}_1}(x_2'), \overline{\mu}_{\widetilde{A}_2}(x_2') \right]$. Two firing levels are then

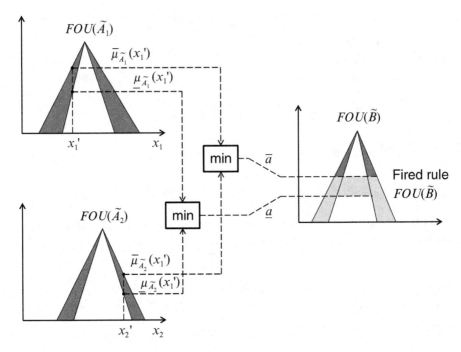

Fig. 1.13 Reasoning in IT2FLS

computed, a lower firing level, $\underline{a}(x')$, and an upper firing level, $\overline{a}(x')$, where $\underline{a}(x')$ $= \min\left[\mu_{\underline{A}_1}(x_1'), \mu_{\underline{A}_2}(x_2')\right]$ and $\overline{a}(x') = \min\left[\overline{\mu}_{\widetilde{A}_1}(x_1'), \overline{\mu}_{\widetilde{A}_2}(x_2')\right]$. $\underline{a}(x')$ is then T-normed with LMF $\left(\widetilde{B}\right)$ and $\overline{a}(x')$ is then T-normed with UMF $\left(\widetilde{B}\right)$. Using the minimum T-norm the resulting fired-rule FOU is the trapezoidal FOU, shown in Fig. 1.13.

References

Aliev RA (1995) Fuzzy knowledge based Intelligent robots, Moscow: Radio i svyaz, (in Russian).

Aliev RA, Aliev RR (1997–1998) Soft computing, Baku: ASOA Press, (in Russian), vol I, II, III

Aliev RA., Aliev RR. (2001) Soft computing and its application. World Scientific, New Jersey, London, Singapore, Hong Kong.

Aliev RA, Tserkovny A (1988) The knowledge representation in intelligent robots based on fuzzy sets, Soviet Math. Doklady, 37, 541–544.

Aliev RA, Mamedova GA, Tserkovny AE (1991a) Fuzzy control systems. Moscow: Energoatomizdat

Aliev RA., Aliev FT, Babaev MD (1991b) Fuzzy process control and knowledge engineering. Koln: Verlag TUV Rheinland

Aliev RA, Tserkovny AE (2011) A systemic approach to fuzzy logic formalization for approximate reasoning. Information Sciences, 181, 1045–1059

Aliev RA, Mamedova GA, Tserkovny AE (1991) Fuzzy control systems. Moscow: Energoatomizdat

Aliev RA., Aliev FT, Babaev MD (1991) Fuzzy process control and knowledge engineering. Koln: Verlag TUV Rheinland

Aliev RA, Bonfig KW, Aliev FT (1993) Messen, Steuern und Regeln mit Fuzzy Logik. München: Franzis-Verlag, (in German)

Aliev RA, Fazlollahi B, Aliev RR (2004) Soft computing and its application in business and economics. Berlin, Heidelberg: Springer-Verlag

Baldwin J F, Pilsworth BW (1979) A model of fuzzy reasoning through multivalued logic and set theory. Int. J. Man-Machines Studies, 11, 351–380.

Bandler W, Kohout L (1980) Fuzzy power sets and fuzzy implications operators. Fuzzy Sets and Systems, 1, 13–30.

Castillo O (2011) Type-2 fuzzy logic in intelligent control applications. Springer, USA.

Castillo O, Melin P (2008) Type-2 fuzzy logic: Theory and applications. Springer, Berlin.

Castillo O, Melin P (2008b) Chapter 1 Introduction to type-2 fuzzy logic. In: Castillo O, Melin P (eds) Type-2 Fuzzy Logic: Theory and Applications. Studies in Fuzziness and Soft Computing, Springer, Berlin.

Dinagar SD, Anbalagan A (2011) Fuzzy programming based on type-2 generalized fuzzy numbers. International J. of Math. Sci. and Eng. Appls., 5, No. 4, 317–329.

Dinagar SD, Latha K (2012) A note on type-2 triangular fuzzy matrices. International J. of Math. Sci. and Eng. Appls., 6, No. 1, 207–216.

Dubois D, Prade H (1980) Fuzzy Sets and Systems: Theory and Applications. NY: Academic Press.

Fan Z-P, Feng B (2009) A multiple attributes decision making method using individual and collaborative attribute data in a fuzzy environment. Information Sciences, 179, 3603–3618.

Fukami S, Mizumoto M, Tanaka K (1980) Some considerations of fuzzy conditional inference. Fuzzy Sets and Systems, 4, 243–273.

Hu Q, Yu D, Guo M (2010) Fuzzy preference based rough sets. Information Sciences, 180, Special Issue on Intelligent Distributed Information Systems, 15, 2003–2022.

Jaffal H, Tao C (2011) Multiple attributes group decision making by type-2 fuzzy sets and systems. Blekinge Institute of Technology, Master Degree Thesis no: 2011:1.

Karnik NN, Mendel JM (1998) An introduction to type-2 fuzzy logic systems. USC, Rep., http://sipi.usc.edu/˜mendel/report.

Karnik NN, Mendel JM (1998b) Introduction to type-2 fuzzy logic systems. Presented at the 1998 I.E. FUZZ Conf. Anchorage, AK.

Karnik NN, Mendel JM (2001) Operations on type-2 fuzzy sets. Fuzzy Sets Syst., 122, 327–348.

Karnik NN, Mendel JM,. Liang Q (1999) Type-2 fuzzy logic systems. IEEE Trans. Fuzzy Systems 7 (6), 643–658.

Kaufman A (1973) Introduction to theory of fuzzy sets. Vol. 1, Orlando: Academic Press.

Klir GJ, Yuan B (1995) Fuzzy sets and fuzzy logic. Theory and Applications, NJ: PRT Prentice Hall .

Klir GJ, Clair US, Yuan B (1997) Fuzzy Set Theory, Foundations and Applications. NJ: PTR Prentice Hall.

Li D-F, Chen GH, Huang ZG (2010) Linear programming method for multiattribute group decision making using IF sets. Information Sciences, 180, 1591–1609.

Liang Q and Mendel JM (2000) Interval type-2 fuzzy logic systems: theory and design. IEEE Trans. Fuzzy Systems, Vol.8, No.5.

Mamdani EH (1977) Application of fuzzy logic to approximate reasoning using linguistic syntheses. IEEE Transactions on Computers, C-26(12), 1182–1191.

Medina J, Ojeda-Aciego M (2010) Multi-adjoint t-concept lattices. Information Sciences, 180, 712–725.

Mendel JM (2001) Uncertain rule-based fuzzy logic systems: Introduction and New Directions, Prentice-Hall.

Mendel JM (2003) Fuzzy sets for words: a new beginning. IEEE FUZZ Conference.

Mendel JM (2007) Type-2 fuzzy sets and systems: an overview. IEEE Computational Intelligence Magazine, vol. 2.

Mendel JM (2008), Chapter 25. On type–2 fuzzy sets as granular models for words. In: Pedrycz W, Skowron A, Kreinovich V (eds) Handbook of Granular Computing, Wiley, England.

Mendel JM, John RIB (2002) Type-2 fuzzy sets made simple. IEEE Trans. Fuzzy Syst., vol. 10, No. 2, 117–127.

Mendel JM, John RI, Liu F (2006) Interval type-2 fuzzy logic systems made simple. IEEE Transactions on Fuzzy Systems, vol. 14, No.6, 808–821

Mizumoto M, Zimmermann H-J (1982) Comparison of fuzzy reasoning methods. Fuzzy Sets and Systems, 8, 253–283.

Mizumoto M, Fukami S, Tanaka K (1979) Some methods of fuzzy reasoning. In: R. Gupta, R. Yager (eds.), Advances in Fuzzy Set Theory Applications, (North-Holland, New York).

Utkin LV (2007) Risk analysis and decision making under incomplete information. St. Petersburg: Nauka, 2007 (in Russian).

Yeh RT, Bang SY (1975) Fuzzy relations, fuzzy graphs, and their applications to clustering analysis. In: Zadeh LA., Fu KS., and Shimura MA. (eds) Fuzzy Sets and Their Applications. NY Academic Press, 125–149.

Zadeh LA (1965) Fuzzy sets. Information and Control, 8, 338–353.

Zadeh LA (1971) Similarity relations and fuzzy orderings. Information Sciences, 3, 177–200.

Zadeh LA (1973) Outline of a new approach to the analysis of complex system and decision processes. IEEE Trans. Systems, Man, and Cybernetics 3, 28–44.

Zadeh LA (1975) The concept of a linguistic variable and its applications in approximate reasoning. Information Sciences, 8, 43–80, pp. 301–357; 9, pp. 199–251.

Zadeh LA (1978) Fuzzy sets as a basis for a theory of possibility. Fuzzy Sets and Systems, 1, 3–28

Zadeh LA (1988) Fuzzy logic. IEEE Computer, 21 (4), 83–93.

Zadeh LA (2005) Toward a generalized theory of uncertainty — an outline. Information Sciences, 172, 1–40.

Zadeh LA (2008) Is there a need for fuzzy logic? Information Sciences, 178, 2751–2779.

Zamri N, Abdullah L, Hitam MS, Noor M, Maizura N, Jusoh A (2013) A novel hybrid fuzzy weighted average for MCDM with interval triangular type-2 fuzzy sets. WSEAS Transactions on Systems, Issue 4, vol.12, 212–228.

Zimmermann H-J (1996) Fuzzy set theory and its applications. Norwell, MA, USA: Kluwer Academic Publishers

Chapter 2
Evolutionary Computing and Type-2 Fuzzy Neural Networks

2.1 Evolutionary Computing Methods

Evolutionary computing involves stochastic search and continuous optimization methods that are inspired by biological mechanisms and systems (Crosby 1973; Eiben and Smith 2003). These computing methods inherit the principles of development and progress from the natural processes and phenomena such as evolution, reproduction (or generation), selection, survival, grouped and distributed behavior, chance, inheritance, crossover (or recombination), mutation, fitness (or health) and so on. Evolutionary computing methods emulate the laws of natural evolution such as "a stronger (healthier or fitter) organism has more chances to survive than a weaker one", "an organism or a pair can generate a new offspring with a probability", "an offspring takes over some of properties of their parents", "an offspring very rarely but may have some properties that differ it from their parents", "population size cannot grow infinitely", etc. Some less natural laws can exist as well: "the best organism will never die".

Very often evolutionary computing based methods are also named population based. This is because the notion of "population (of individuals)" forms the basis and exists in all such methods whereas the type (i.e. its design and set of properties) of individuals and of the population as well as the processing algorithms to evolve the population may vary in a wide range. In all evolutionary computing methods, every individual is attached a numerical value reflecting its fitness (quality, healthiness). There should also be provided a way to derive the individual's fitness degree from its properties.

Well-known evolutionary computing techniques are: Differential Evolution (Storn and Price 1997; Price et al. 2005; Feoktistov 2006), Swarm Optimization (Bonabeau et al. 1999; Clerc 2006; Kennedy and Eberhart 1995), Ant Colony Optimization (Dorigo and Stützle 2004), Cultural Algorithms (Reynolds 1994), Harmony Search (Geem et al. 2001; Karahan et al. 2012; Ricart et al. 2011),

© Springer International Publishing Switzerland 2014
R.A. Aliev, B.G. Guirimov, *Type-2 Fuzzy Neural Networks and Their Applications*,
DOI 10.1007/978-3-319-09072-6_2

Genetic Algorithm (Chiong et al. 2012; Goldberg 1989; Langdon and Poli 2002) and others.

When applying an evolutionary computing based approach for solving optimization problems a candidate solution (i.e. appropriate values of sought-for variables) is represented as an individual in a population and the corresponding value of objective function (possibly normalized) as the individual's fitness degree. Generation of new individuals (and accordingly the candidate solutions), their survival, and overall treatment of the population are governed by the laws of evolution driven by application of multiple so-called *evolutionary forces* (or operators) implemented within a specific evolutionary computing technique. The most important and frequently used evolutionary forces are recombination (crossover), mutation, selection, and elitism. While recombination and mutation creates diversity in the population (and accordingly, in the candidate solutions), selection and elitism increases its quality. Thereby it is implemented a global and continuous optimization process.

In some techniques such as Genetic Algorithms (GA) it may be required some transformation procedure to get the problem's decision or search variables into individuals and back (encoding/decoding). Or more specifically: the procedure to convert the variables' numerical values into instances of the individual's *container class* (i.e. a specific data type with fields representing an individual's properties) and vice versa. In GA they often call such data containers as chromosomes or genes (genomes). Physically, in computers, the chromosomes are represented as long strings of bits. As they express it in genetic algorithms, the phenotype (numeric values of problem's variables) is encoded to produce a genotype (a data container – gene or chromosome).

The assessment of the individuals is done by a function called a fitness function (GA) or by a computational model that allows computation of the individual's fitness degree from its properties (phenotype or genotype). If the algorithm uses data containers such as chromosomes, at some stage of the evolution the best (healthiest, strongest, or fittest) chromosome should have been decoded to retrieve the corresponding values of decision variables (e.g. the sought-for solution).

For optimization problems solved by evolutionary computing methods, the fitness function (sometimes, in such techniques as Differential Evolution – DE – replaced by cost or error function) is produced from the objective function and, possibly, constraints posed on the decision variables.

Genetic Algorithm (GA) is one of the first offered evolutionary computing methods and is still very popular. Figure 2.1 in a very general form illustrates the scheme of GA.

Please notice the force named elitism that we include in the scheme of GA shown in Fig. 2.1. The *elitism* force ensures that at least one of best chromosomes is transferred to the next generation. The *elitism* force is not an absolutely necessary one and does not exist in basic versions of GA. However, it is very useful to guarantee the best ever reached historical solution is never lost.

Figure 2.2 illustrates a possible version of implementation of the Genetic Algorithm.

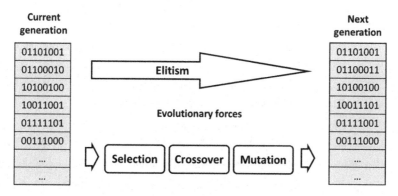

Fig. 2.1 The genetic algorithm

This version of GA requires five user defined parameters: population size (*PopSize*), probability of crossover (*CrossoverProb*), probability of mutation (*MutationProb*), maximum number of generations (*MaxGen*), and desired fitness (*FitnessDesired*). It is assumed that there exist functions: (1) to produce the chromosome from a candidate solution (*get_genotype*), (2) to produce the candidate solution from a chromosome (*get_phenotype*), and (3) to evaluate the quality (fitness degree) of a candidate solution (*fitness*). Also it is assumed that there are operators implementing evolutionary forces (*crossover* and *mutate*). The *crossover* force takes two chromosomes as its arguments to produce a new offspring chromosome, and the *mutate* force alters its single argument chromosome. An i-th chromosome is designated *Chromosome*[i], $i = 1, \ldots$, *PopSize*. The algorithm stops when the desired fitness threshold (*FitnessDesired*) is reached by one of the chromosomes or the maximum generation number (*MaxGens*) is passed.

As can be seen from the algorithm presented in Fig. 2.2, at steps 3.1–3.4 the bunch of evolutionary forces *elitism*, *selection*, *crossover* (recombination), and *mutation* are applied to the chromosomes in the existing population to create members of the new population. After meeting the stop condition, the sought-for solution is extracted from the chromosome with maximum fitness.

The versions of GA may differ from not only the set but also the type of evolutionary forces (genetic operators) used. Figure 2.3 illustrates various implementations of the crossover and mutation forces.

Let's now consider an example of optimization using the GA (Fig. 2.4).

Example Find the maximum of the function:

$$z = f(x, y) = 3(1 - x)^2 e^{-x^2 - (y+1)^2} - 10\left(\frac{x}{5} - x^3 - y^5\right) e^{-x^2 - y^2} - \frac{1}{3} e^{-(x+1)^2 - y^2}$$

Figure 2.5 demonstrates how the initial 20 candidate solutions are distributed and how the population changes after five and ten generations.

GA(*PopSize, SelectionProb, MutationProb, MaxGen, FitnessDesired*)

Step 1.   Create and initialize *Population* with *PopSize* random chromosomes: *Ch[i]*, $i=(1,..,PopSize)$. *Gen*=0

Step 2.   Evaluate fitness for each chromosome in *Population*: *Fitness[i]=fitness(get_phenotype(Cromosome[i]))*

Step 3.   Create new generation *New Population*:

      Step 3.1.   Apply *elitism*: copy the best *Cromosom* $[\arg\max_i(Fitness[i])]$

                      from *Population* to *NewPopulation*.

      Step 3.2.   Select two chromosomes cr_k and cr_l from *Population* with probabilities $\dfrac{Fitness[k]}{\sum_j Fitness[j]}$ and $\dfrac{Fitness[l]}{\sum_j Fitness[j]}$, respectively.

      Step 3.3.   Apply crossover force for the selected pare (cr_k, cr_l) with *CrossoverProb* to produce an offspring: $cr_{new} = crossover(cr_k, cr_l)$

      Step 3.4.   Apply mutation force with *Mutation Prob*: $cr_{new} = mutate(cr_{new})$

      Step 3.5.   Copy cr_{new} to *NewPopulation*.

      Step 3.6.   If number of chromosomes in *New Population* < *PopSize* go to Step3.2.

Step 4.   Replace *Population* with *NewPopulation*. Increment *Gen=Gen*+1

Step 5.   Extract the solution: *Solution=get_phenotype(Cromosone[* $\arg\max_i(Fitness[i])$ *])*

Step 6.   Check the termination condition: If $\left(\max_i(Fitness[i]) \geq FitnessDesired\right)$ or $(Gen \geq MaxGens)$ terminate and return *Solution*, else go to Step2

Fig. 2.2 Genetic algorithm implementation high level code

Particle swarm optimization (PSO) is another population based stochastic optimization technique inspired by the social behavior of birds (Kennedy and Eberhart 1995, 2001). The algorithm is very simple but powerful. The PSO algorithm is quite similar to genetic algorithms and can be used for similar problems. Because the algorithm allows for parallelization and uses less computational resources than GA, it can efficiently be used to minimize/maximize high dimensional functions and thus as a training method for neural networks.

To understand the algorithm, it is best to imagine a swarm of birds that are searching for food in a defined area. It is assumed that there is only one piece of food in this area. Initially, the birds don't know where the food is, but they know at

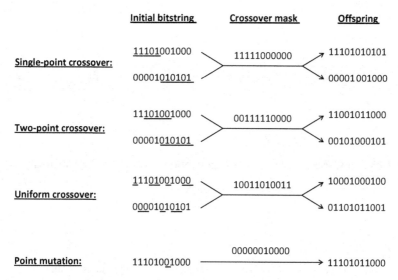

	Initial bitstring	Crossover mask	Offspring

Single-point crossover:

11101001000   11111000000   11101010101
00001010101   00001001000

Two-point crossover:

11101001000   00111110000   11001011000
00001010101   00101000101

Uniform crossover:

11101001000   10011010011   10001000100
00001010101   01101011001

Point mutation:

                   00000010000
11101001000   ─────────────────→ 11101011000

Fig. 2.3 Implementation of operators for GA

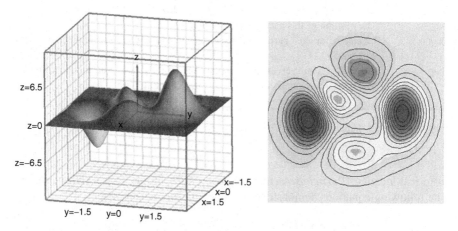

Fig. 2.4 Example function to maximize using GA

each time how far the food is. It is quite natural that in their search the birds will follow the strategy that is nearest to the food.

PSO adopts this behavior and searches the search space for candidate solutions, which are called here particles. Very similar to the GA, each particle has its cost degree (fitness) that is evaluated by a function to be minimized, and each particle has a velocity that directs its flying.

The swarm is initialized by particles at random positions and then each particle flies through the search space by adjusting its velocity (vector **V**) and location (vector **X**) to follow two best candidate solutions found so far: their own best

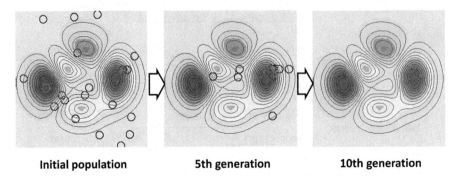

Initial population   **5th generation**   **10th generation**

Fig. 2.5 The process of optimization of the example function by GA

PSO(*PopSize*, c_1, c_2, *MaxGen*, *MinimumCost*)

Step 1.	Initialize each particle with a random velocity and random position.
Step 2.	Calculate the cost for each particle. If the current cost is lower than the best value so far, remember this position (*ParticleBest*).
Step 3.	Choose the particle with the highest fitness (lowest cost) of all particles. The position of this particle is *GlobalBest*.
Step 4.	Calculate, for each particle, the new velocity and position according to the equations (2.1).
Step 5.	Go to Step 2 and repeat steps 2-4 until one of the criteria (maximum iterations *Max Gen* or minimum cost *Minimum Cost*) is not attained.

Fig. 2.6 Basic PSO algorithm

historical location (*ParticleBest*) and the best particle in the swarm (*GlobalBest*). The following equations demonstrate how the adjustments are made:

$$\mathbf{V}_{new} = \mathbf{V} + c_1 r_1 (ParticleBest - \mathbf{X}) + c_2 r_2 (GlobalBest - \mathbf{X})$$
$$\mathbf{X}_{new} = \mathbf{X} + \mathbf{V}, \qquad (2.1)$$

where $\mathbf{V}$ is the current velocity, $\mathbf{V}_{new}$ is the new velocity, $\mathbf{X}$ is the current position, $\mathbf{X}_{new}$ is the new position, r_1 and r_2 are random numbers in the interval [0, 1], and c_1 and c_2 are acceleration coefficients: c_1 is the factor that influences the cognitive behavior, i.e. how much the particle will follow its own best solution, and c_2 is the factor for social behavior, i.e. how much the particle will follow the swarm's best solution.

In an optimization problem, the current position will be meant as the vector of sought-for parameter values, i.e. a candidate solution.

The basic algorithm can be written as presented in Fig. 2.6:

The next section will consider another evolutionary optimization technique, which is very efficient and suitable, in our opinion, for application in training of parameters of neural networks – the differential evolution (DE) algorithm.

2.2 Differential Evolution Based Optimization (DEO)

In this section we consider the Differential Evolution (DE) algorithm. This population based algorithm implements global search. Being designed specifically for numerical optimization, it is characterized by good convergence properties in multidimensional search spaces. DE has been successfully applied to solve a wide range of problems such as those found in image classification, clustering, and function optimization. These characteristics of DE make the method also an efficient tool for implementation of neural network training.

2.2.1 DE Algorithm

As other stochastic and population-based methods, DE algorithm (Price et al. 2005; Storn and Price 1997; Chakraborty 2008; Feoktistov 2006) uses an initial population of randomly generated individuals and applies to them operations of differential mutation, crossover, and selection. DE considers individuals as vectors in n-dimensional Euclidean space. The population of $PopSize$ ($PopSize \geq 4$) individuals is maintained through consecutive generations. A new vector is generated by mutation, which, in this case is completed by adding a weighted difference vector of two individuals to a third individual as follows: $\mathbf{X}_{new} = (\mathbf{X}_{r1} - \mathbf{X}_{r2})f + \mathbf{X}_{r3}$, where $\mathbf{X}_{r1}, \mathbf{X}_{r2}, \mathbf{X}_{r3}$ ($r_1 \neq r_2 \neq r_3$) are three different individuals randomly picked from the population and f (>0) is the mutation parameter. The mutated vector then undergoes crossover with another vector thus generating a new offspring.

The selection process is realized as follows. If the resulting vector is better (e.g. yields a lower value of the cost function) than the member of the population with an index changing consequently, the newly generated vector will replace the vector with which it was compared in the following generation. Another approach, which we adopted in this research, is to randomly pick an existing vector for realizing crossover.

Figure 2.7 illustrates a process of generation of a new trial solution (vector) $\mathbf{X}_{new}$ from three randomly selected members of the population $\mathbf{X}_{r1}, \mathbf{X}_{r2}, \mathbf{X}_{r3}$. Vector $\mathbf{X}_i$, $i = 1, \ldots, PopSize$, $i \neq r_1 \neq r_2 \neq r_3$ becomes the candidate for replacement by the new vector, if the former is better in terms of the DE cost function. Here, for illustrative purposes, we assume that the solution vectors are of dimension $n = 2$ (i.e. two parameters are to be optimized).

The algorithm itself can be described as presented in Fig. 2.8.

Fig. 2.7 Realization of DE optimization: a two-dimensional case

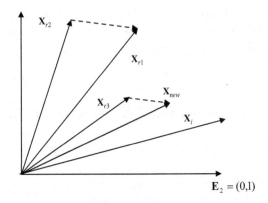

$$\mathbf{E}_2 = (0,1)$$

DEO(*PopSize, f, cr, MaxGen, MinimumCost*)

Step 1.   Randomly generate *PopSize* parameter vectors (from respective parameter spaces (e.g. in the range [-1, 1]) and form a population $P=\{\mathbf{X}_1, \mathbf{X}_2, ..., \mathbf{X}_{ps}\}$

Step 2.   While the termination condition (maximum iterations *MaxGen* reached or minimum cost *MinimumCost* attained) is not met generate new parameter sets:

      Step 2.1.   Choose a next vector $\mathbf{X}_i$ (*i*=1,...,*PopSize*)

      Step 2.2.   Choose randomly different 3 vectors from *P*: $\mathbf{X}_{r1}, \mathbf{X}_{r2}, \mathbf{X}_{r3}$ each of which is different from current X_i

      Step 2.3.   Generate trial vector $\mathbf{X}_t = \mathbf{X}_{r1} + f(\mathbf{X}_{r2} - \mathbf{X}_{r3})$

      Step 2.4.   Generate a new vector from trial vector $\mathbf{X}_t$. Individual vector parameters of $\mathbf{X}_t$ are inherited with probability *cr* into the new vector $\mathbf{X}_{new}$. If $\mathbf{X}_{new}$ evaluates as being a better solution than $\mathbf{X}_i$, then the current $\mathbf{X}_i$ is replaced in population *P* by $\mathbf{X}_{new}$

     Next i

Step 3.   Select from population *P* the parameter vector $\mathbf{X}_{best}$, which is evaluated as the best solution

Step 4.   Stop the algorithm

Fig. 2.8 Basic DE algorithm

Usually the mutation rate f is chosen $f \in [0, 2]$. After the crossover of the trial vector and the vector $\mathbf{X}_i = (x_i[1], x_i[2], ..., x_i[n])$ from the population, at least one of elements (dimensions) of the trial vector $\mathbf{X}_t = (x_t[1], x_t[2], ..., x_t[n])$ should be transferred to the offspring vector $\mathbf{X}_{new} = (x_{new}[1], x_{new}[2], ..., x_{new}[n])$. The crossover parameter $cr \in [0, 1]$ affects the mutated (trial) vector as follows:

$$x_{new}[j] = \begin{cases} x_{new}[j], & \text{if } rand(0, 1) \le cr \text{ or } rand(1, n) = j, \\ x_i[j], & \text{otherwise,} \end{cases}$$

where the function *rand(a,b)* returns a random value in the range [a,b].

Fig. 2.9 The Rosenbrock
function

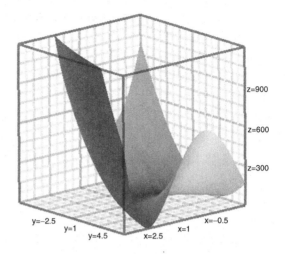

Differential Evolution based Optimization (DEO) usually implies existence of a function $F(\mathbf{X})$, where $\mathbf{X}$ is vector $(x_1, \ldots, x_n)$, whose values should be minimized by DE algorithm. In DE based neural network training $F(\mathbf{X})$ is replaced by the network error function.

When solving optimization problems a common recommendation is to choose *PopSize* ten times the number of optimization variables (Price et al. 2005), $f = 0.9$, $cr = 1$ or $cr = 0.5$. Some authors suggest ways for choosing optimal values of DE parameters f and cr. For example (Brest et al. 2006) suggests a way for self-adapting of the DE parameters during the optimization.

Let's consider a couple of examples of function optimization using DE.

Example Find the minimum of the function (the Rosenbrock function): $z(x, y) = 100(x^2 - y)^2 + (x - 1)^2$, see Fig. 2.9.

Using the code implementing the DE algorithm, we can see that the algorithm very quickly finds the minimum of this function ($z_{min} = 0$ reached at $(x_{min}, y_{min}) = (1, 1)$) In Fig. 2.10 you can see the average number of calculations of the function to reach the minimum for specified error levels. The experiment has been done with the standard version of DE with the parameters set as *PopSize* $= 10$, $f = 0.9$, $cr = 1$.

DE is also very effective for multi-parametric function optimization and outperforms most classical and other evolutionary algorithms in finding global minimums for non-smooth and multi-extreme functions. The suggested standard version of the DE algorithm (with *PopSize* set to 3,000) used for minimization of the 50-dimensional Rosenbrock function has used on average about 12 million function evaluations to reach accuracy of 10^{-6}. This performance is comparable with gradient-based methods. However, it should be noted that the well-known coordinate decent, gradient based, and many other classical methods are not global optimizers (the Rosenbrock function is indeed a single-extreme function) and

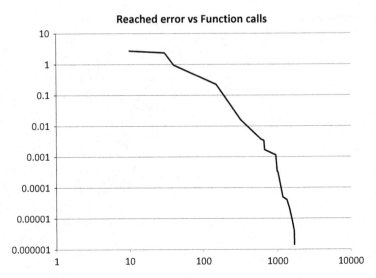

Fig. 2.10 Performance of DE being used for minimization of the Rosenbrock function

pose certain requirements on the function such as smoothness, continuity, differentiability, etc.

Let's consider another example with a more complex function with many local minima.

Example Find the minimum of Griewangk's function (Griewangk 1981):

$$f(\mathbf{X}) = \sum_{i=0}^{9} \frac{x_i^2}{4000} - \prod_{i=0}^{9} \cos\left(\frac{x_i}{\sqrt{i+1}}\right) + 1, \, x_i \in [-400, 400].$$

Its global minimum $f(\mathbf{0}) = 0$ is very difficult to find. The standard DE version (*PopSize* = 50) has used on average about 32,000 function evaluations to find the global minimum with the accuracy 10^{-6}.

The Tables 2.2 and 2.3 compares the performance of DE and other methods on a number of benchmark functions presented in Table 2.1 (Vesterstrøm and Thomsen 2004; Yao and Liu 1996).

For DE the following control parameters were set: *PopSize* = 100, $cr = 0.9$, $f = 0.5$. For PSO: *PopSize* = 25, $c_1 = c_2 = 1.8$. For GA: *Popsize* = 100, *MutationProb* = 0.9, *CrossoverProb* = 0.7.

For each problem 30 runs of each algorithm were done and the mean value of function minimum reached and the standard deviation were computed. In Tables 2.2 and 2.3 the best performing algorithms for each problem are marked in bold.

As can be seen from the results of experiments presented above (Vesterstrøm and Thomsen 2004), the performance of DE is outstanding in comparison to the other evolutionary algorithms tested. DE has found the optimum in almost every run.

Table 2.1 Numerical benchmark problems

#	Function	Dimension	Ranges	Minimum
1	$\sum_{i=0}^{n-1} x_i^2$	30/100	$[-5.12, 5.12]$	$F(\mathbf{0}) = 0$
2	$\sum_{i=0}^{n-1} \lvert x_i \rvert + \prod_{i=0}^{n-1} x_i$	30/100	$[-10,10]$	$F(\mathbf{0}) = 0$
3	$\sum_{i=0}^{n-1} \left(\sum_{j=0}^{i} x_i \right)^2$	30/100	$[-100,100]$	$F(\mathbf{0}) = 0$
4	$\max \lvert x_i \rvert,\ i = 0,\ldots,n-1$	30/100	$[-100,100]$	$F(\mathbf{0}) = 0$
5	$\sum_{i=0}^{n-1} (100(x_{i+1} - x_i^2)^2 + (x_i - 1)^2)$	30/100	$[-30,30]$	$F(\mathbf{1}) = 0$
6	$\sum_{i=0}^{n-1} \left(\left\lfloor \left\lvert x_i + \frac{1}{2} \right\rvert \right\rfloor \right)^2$	30/100	$[-100,100]$	$F(\mathbf{p}) = 0,$ $0.5 \le p_i < 0.5$
7	$\left(\sum_{i=0}^{n-1} (i+1) x_i^4 \right) + rand(0,1)$	30/100	$[-1.28,1.28]$	$F(\mathbf{0}) = 0$
8	$\sum_{i=0}^{n-1} \left(-x_i \sin\left(\sqrt{\lvert x_i \rvert} \right) \right)$	30/100	$[-500,500]$	$F(\mathbf{420.97}) = -12569.5/-41898.3$
9	$\sum_{i=0}^{n-1} (x_i^2 - 10\cos(2\pi x_i) + 10)$	30/100	$[-5.12, 5.12]$	$F(\mathbf{0}) = 0$
10	$-20\exp\left(-0.2\sqrt{\frac{1}{n} \sum_{i=0}^{n-1} x_i^2} \right) - \exp\left(\frac{1}{n} \sum_{i=0}^{n-1} \cos(2\pi x_i) \right) + 20 + e$	30/100	$[-32,32]$	$F(\mathbf{0}) = 0$
11	$\frac{1}{4000}\left(\sum_{i=0}^{n-1} x_i^2 \right) + \prod_{i=0}^{n-1} \cos\left(\frac{x_i}{\sqrt{i+1}} \right) + 1$	30/100	$[-600,600]$	$F(\mathbf{0}) = 0$

(continued)

Table 2.1 (continued)

#	Function	Dimension	Ranges	Minimum
12	$\dfrac{\pi}{n}\Big\{10(\sin(\pi y_1))^2 +$ $\sum_{i=0}^{n-2}\Big((y_i-1)^2\big(1+10(\sin(\pi y_{i+1}))^2\big)\Big) +$ $(y_n-1)^2\Big\} + \sum_{i=0}^{n-1} u(x_i, 10, 100, 4),$ where : $y_i = 1 + \dfrac{1}{4}(x_i+1),$ $u(x,a,b,c) = \begin{cases} b(x-a)^c, & \text{if } x > a \\ b(-x-a)^c, & \text{if } x < -a \\ 0, & \text{if } -a \le x \le a \end{cases}$	30/100	$[-50,50]$	$F(-\mathbf{1}) = 0$
13	$\sum_{i=0}^{10}\left(a_i - \dfrac{x_0\big(b_i^2+b_i x_1\big)}{b_i^2+b_i x_2+x_3}\right)^2,$ where : $a = (0.1957, 0.1947, 0.1735, 0.1600, 0.0844,$ $0.0627, 0.0456, 0.0342, 0.0323, 0.0235, 0.0246)$ $b = \left(4, 2, 1, 0.5, 0.25, \tfrac{1}{6}, \tfrac{1}{8}, \tfrac{1}{10}, \tfrac{1}{12}, \tfrac{1}{14}, \tfrac{1}{16}\right)$	4	$[-5,5]$	$F(0.19, 0.19, 0.12, 0.14) = 0.0003075$
14	$4x_0^2 - 2.1x_0^4 + \tfrac{1}{3}x_0^6 + x_0 x_1 - 4x_1^2 + 4x_1^4$	2	$[-5,5]$	$F(-0.09, 0.71) = -1.0316$

Table 2.2 Results of evolutionary algorithms on the benchmark problems of dimensionality 30 or less

#	DE		PSO		Standard EA (GA)	
	Mean	Std dev	Mean	Std dev	Mean	Std dev
1	**0.00**	0.00	**0.00**	0.00	1.79E-03	2.77E-04
2	**0.00**	0.00	**0.00**	0.00	1.72E-02	1.70E-03
3	2.02E-09	8.26E-10	**0.00**	0.00	1.59E-02	4.25E-3
4	3.85E-08	9.17E-09	**2.10E-16**	8.01E-16	1.98E-02	2.07E-03
5	**0.00**	0.00	4.03E+00	4.99E00	3.13E+01	1.74E+01
6	**0.00**	0.00	4.00E-02	1.98E-01	**0.00**	0.00
7	4.94E-03	1.13E-03	1.91E-03	1.14E-03	**7.11E-04**	3.27E-04
8	**−1.2569E+04**	2.30E-04	−7.19E+0.3	6.72E+0.2	−1.17E+04	2.34E+02
9	**0.00**	0.00	4.92E+01	1.62E+01	7.18E-01	9.22E-01
10	**−1.19E-15**	7.03E-16	1.40E+00	7.91E-01	1.05E-02	9.08E-04
11	**0.00**	0.00	2.35E-02	3.5E-02	4.64E-03	3.96E-03
12	**0.00**	0.00	3.82	8.40E-01	4.56E-06	8.11E-07
13	4.17E-04	3.01E-04	1.34E-03	3.94E-03	**3.70E-04**	8.78E-05
14	−1.03E00	1.92E-08	−1.03E00	3.84E-08	**−1.03E00**	3.16E-08

Table 2.3 Results of evolutionary algorithms on the benchmark problems of dimensionality 100

#	DE		PSO		Standard EA (GA)	
	Mean	Std dev	Mean	Std dev	Mean	Std dev
1	**0.00**	0.00	**0.00**	0.00	5.23E-04	5.18E-05
2	**0.00**	0.00	1.80E+01	6.52E+01	1.74E-02	9.43E-04
3	**5.87E-10**	1.83E-10	3.67E+03	6.94E+03	3.68E-02	6.06E-03
4	**1.13E-09**	1.42E-10	5.31E+00	8.63E-01	7.67E-03	5.71E-04
5	**0.00**	0.00	2.02E+02	7.66E+02	9.25E+01	1.29E+01
6	**0.00**	0.00	2.10E+00	3.52E+00	**0.00**	0.00
7	7.66E-03	6.58E-04	2.78E-02	7.31E-02	**7.05E-04**	9.70E-05
8	**−4.1898E+04**	1.06E-03	−2.16EE+04	1.73E+03	−3.94E+04	5.36E+02
9	**0.00**	0.00	2.43E+02	4.03E+01	9.98E-02	3.04E-01
10	**8.02E-15**	1.74E-15	4.49E+00	1.73E+00	2.93E-03	1.47E-04
11	**5.42E-20**	0.00	4.17E-01	6.45E-01	1.89E-03	4.42E-03
12	**0.00**	0.00	1.18E-01	1.75E-01	2.98E-07	2.76E-08

2.2.2 Using Constraints in DEO

The constraints can be used in neural network training where specific requirements are posed to neurons parameters' ranges.

Consider we have an optimization problem with constraints:

Minimize $F(\mathbf{X})$
Subject to:

$$G_j(\mathbf{X}) \le 0, \quad j = 1, \ldots, q,$$
$$H_j(\mathbf{X}) = 0, \quad j = q + 1, \ldots, m,$$
$$l_i \le x_i \le u_i, \quad i = 1, \ldots, n,$$

where $\mathbf{X} = (x_1, \ldots, x_n)$, $F(\mathbf{X})$ is the objective function, $G_j(\mathbf{X})$ and $H_j(\mathbf{X})$ are the constraint functions, each variable x_i are limited by lower l_i and upper bounds u_i.

Then we can define a constraints violation function $C(\mathbf{X})$ as follows (Takahama and Sakai 2012):

$$C(\mathbf{X}) = \max\left\{ \max_j \left\{0, G_j(\mathbf{X})\right\}, \ \max\left|H_j(\mathbf{X})\right| \right\}$$

or

$$C(\mathbf{X}) = \sum_{j=1}^{q} \left(\max_j \left\{0, G_j(\mathbf{X})\right\} \right)^2 + \sum_{j=q+1}^{m} \left(H_j(\mathbf{X})\right)^2$$

Then of the two vectors $\mathbf{U}$ and $\mathbf{V}$ (i.e. potential solutions, elements of a DE population) from the population P, whether $\mathbf{U}$ is better than $\mathbf{V}$ can be decided as follows:

$$\mathbf{U} \succ \mathbf{V} \Leftrightarrow (F(U) < F(V) \text{ and } |C(U) - C(V)| \le \varepsilon) \text{ or } (C(U) < C(V))$$

where $\varepsilon \ge 0$ is a small value.

A found solution $\mathbf{X}$ is considered feasible if $C(\mathbf{X}) \le \varepsilon$.

As can be seen the goal is to minimize both the constraint violation function and the objective function. Note also that the minimization of the constraint violation function is more important for unfeasible population vectors than the objective function.

2.2.3   Training of All Types of Neural Networks by DEO

As we have already mentioned in previous sections, evolutionary computing based methods are more flexible than classical methods when using for global optimization of functions (Aliev et al. 2009).

As they do not require any restrictive properties for the functions or the computational models to work with, they can be effectively used for training of parameters of ordinary, fuzzy, and fuzzy type-2 neural networks. The models of neural networks, and especially fuzzy and high-order fuzzy networks can be described by complex nonlinear, non-convex, and non-differentiable functions.

Training of a neural network is in fact a procedure to minimize a function evaluating the network error, e.g. mismatch between the network's actual output and the desired output for a given input. The typical error function is described as follows:

$$E = \frac{1}{n \cdot s_y} \sum_{p=1}^{n} \sum_{i=1}^{s_y} \left(y_{pi}^* - y_{pi} \right)^2$$

Here y_{pi}^* is the desired value (target) for output i when we apply input value vector $\mathbf{x}_p$, y_{pi} is the corresponding output of the model output, n is the number of training patterns, and s_y is the number of outputs in the model.

The decision or optimization variables or parameters for a neural network are either its connection weights (perceptron like NN) or fuzzy parameters describing the input and output linguistic terms in fuzzy rules (NN-based fuzzy inference systems). To apply an evolutionary algorithm, in our case, the DE, the whole bunch of these parameters should have been considered as a population individual.

For example, for a perceptron-like recurrent fuzzy neural network (RFNN), we consider a population individual to represent a whole combination of weights $\left(\widetilde{W} = \left\{ \widetilde{w}_{lij} \right\}, \ \widetilde{V} = \left\{ \widetilde{v}_{lij} \right\} \right)$ and biases $\left(\widetilde{\theta} = \left\{ \widetilde{\theta}_{li} \right\} \right)$ (i.e. parameters of RFNN) defining the input/output mapping (Aliev et al. 2009). For a type-2 neural network, considered in Sect. 3.6.2 of this book, each fuzzy term-parameter is itself described by several sub-parameters: the parameters LL, LR, ML, MR, RL, RR for all input terms and the parameters L, ML, MR, R for all outputs terms (Aliev et al. 2011).

The population maintains a number of potential parameter sets defining different network solutions and recognizes one of these solutions to be the best solution. This best solution is the one with minimum training error. After a series of generations, the best solution may converge to a near-optimum solution, which would represent a network performing with the required accuracy.

The detailed DE based algorithm for training of FNN and FRNN has been presented in Sect. 3.5.2 of this book. Section 3.9.1 presents a DE based training for a type-2 neural network, considered in Sect. 3.6.2 of this book.

References

Aliev RA, Guirimov BG, Fazlollahi B, Aliev RR (2009) Evolutionary algorithm-based learning of fuzzy neural networks. Part 2: Recurrent fuzzy neural networks. Fuzzy Sets and Systems archive, Volume 160 Issue 17, 2553–2566.

Aliev RA, Pedrycz W, Guirimov B, Aliev RR, Ilhan U, Babagil M, Mammadli S (2011) Type-2 fuzzy neural networks with fuzzy clustering and differential evolution optimization. Information Sciences, Volume 181 Issue 9, 1591–1608.

Bonabeau E, Dorigo M, Theraulaz G (1999) Swarm intelligence: from natural to artificial systems. Oxford University Press, USA

Brest J, Greiner S, Boskovic B, Mernik M, Zumer V (2006) Self-Adapting Control Parameters in Differential Evolution: A Comparative Study on Numerical Benchmark Problems. IEEE Transactions on Evolutionary Computation, Vol. 10, No 6.

Chakraborty UK (eds) (2008), Advances in Differential Evolution, Springer

Chiong R, Weise T, Michalewicz Z (eds) (2012) Variants of evolutionary algorithms for real-world applications. Springer

Clerc M (2006) Particle swarm optimization. ISTE

Crosby JL (1973) Computer simulation in genetics. John Wiley & Sons, London

Dorigo M, Stützle T (2004) Ant colony optimization. MIT Press

Eiben A, Smith J (2003). Introduction to evolutionary computing. Springer

Feoktistov V (2006) Differential evolution: in search of solutions. Springer

Geem ZW, Kim JH, Loganathan GV (2001) A new heuristic optimization algorithm: harmony search. Simulation 76: 60–68

Goldberg DE (1989) Genetic algorithms in search, optimization and machine learning. Addison Wesley

Griewangk (1981), A.O., Generalized Descent for Global Optimization, JOTA, vol. 34, pp. 11–39.

Karahan H, Gurarslan G, Geem ZW (2012) Parameter estimation of the nonlinear Muskingum flood routing model using a hybrid harmony search algorithm. Journal of Hydrologic Engineering. doi:10.1061/(ASCE)HE.1943-5584.0000608

Kennedy J, Eberhart R (1995) Particle swarm optimization. In: Proceedings of IEEE International Conference on Neural Networks IV: 1942–1948. doi:10.1109/ICNN.1995.488968

Kennedy J, Eberhart R (2001) Swarm intelligence. Morgan Kaufmann Publishers, San Francisco

Langdon WB, Poli R (2002) Foundations of genetic programming. Springer-Verlag

Price K, Storn, RM, Lampinen JA (2005). Differential evolution: a practical approach to global optimization. Springer

Reynolds RG (1994) An introduction to cultural algorithms. In: Proceedings of the 3rd Annual Conference on Evolutionary Programming. World Scientific Publishing: 131–139

Ricart J, Hüttemann G, Lima J, Barán B (2011) Multiobjective harmony search algorithm proposals. Electronic Notes in Theoretical Computer Science

Storn R, Price K (1997) Differential evolution – a simple and efficient heuristic for global optimization over continuous spaces. Journal of Global Optimization 11: 341–359. doi:10.1023/A:1008202821328

Takahama T, Sakai S (2012) Efficient Constrained Optimization by the ε Constrained Rank-Based Differential Evolution. In: Proc. of 2012 I.E. Congress on Evolutionary Computation (CEC).

Vesterstrøm J, Thomsen R (2004) A Comparative Study of Differential Evolution, Particle Swarm Optimization, and Evolutionary Algorithms on Numerical Benchmark Problems. In: Congress on Evolutionary Computation (CEC2004), Volume:2, 1980–1987.

Yao X, Liu Y (1996) Fast evolutionary programming. In: Fogel LJ, Angeline PJ, Back T (eds), Proceedings of the 5th Annual Conference on Evolutionary Programming, 451–460. MIT Press.

Chapter 3
Type-1 and Type-2 Fuzzy Neural Networks

3.1 Introduction to Artificial Neural Networks

An artificial neural network is a computational model simulating a biological neural network (Negnevitsky 2005; Haykin 1999; Ham and Kostanic 2001; Aliev et al. 2004). The latter is indeed a system used by biological organisms to process information from the environment for further decision making. Thus, artificial neural networks are models that mimic the behavior of biological processing systems from accepting data by the nerve endings, processing in the brains, and sending output reactions to other biological systems.

Artificial neural networks (ANN) inherit such structural and functional properties of biological analogs (BANN) as parallel processing (of large number of interconnected simple processing units with different sensitivity of connections), training-by-example ability, time-consuming (and also generally not success-guaranteed) training, generalization ability, robustness, compactness, redundancy, and others (Bullinaria 2013; Aliev et al. 2004).

ANN, similarly to BANN, demonstrates an individual effect obtained from a special organization of simple processing elements inside an integral system. Loss of any simple neuron does not seriously degrade a large neural network (NN). The behavior is determined by the joint effect of large number of distributed and interconnected neurons and the way their connection strengths are tuned up, i.e. by their organization. We can say also that the behavior is determined not much by the quantity but by the quality. Indeed, more intelligent biological species cannot be selected from less intelligent ones just based on their head sizes.

An Artificial Neural Network is specified by:

- a model of information processing unit of NN (artificial neuron),
- architecture of NN, i.e. a collection of neurons and their connection styles and weights (strengths), and
- one or more training (learning) algorithms used to adjust the connection weights of NN in order to model a particular behavior.

© Springer International Publishing Switzerland 2014
R.A. Aliev, B.G. Guirimov, *Type-2 Fuzzy Neural Networks and Their Applications*,
DOI 10.1007/978-3-319-09072-6_3

Whichever model, architecture, and training algorithm is chosen, a network should be able to perform its main task – train and simulate a behavior of an unknown system with any required degree of accuracy.

3.1.1 Neuron Models

Dependent of the neuron's model, networks can be divided into binary, linear, and non-linear ones. The non-linear ones can be further divided into perceptron-like (with a nonlinear, e.g. sigmoidal activation function), radial-function-based, logical, and others. Also according to the way the networks represent input and output signals and parameters they can be crisp, stochastic, interval, fuzzy, type-2 fuzzy and others.

Neurons can be static (ordinary, non-dynamic) or dynamic. Static neurons do not remember their output values (i.e. their historical states), while dynamic neurons store temporary states for two or more periods. The output value of dynamic neurons (with fixed parameters) is time-dependent, i.e. it can depend not only on current input values but also the history of previous output values.

Mathematically, a neuron can be defined as a parameterized function with two or more arguments (inputs) all accepting data of a single predetermined numerical type (e.g. real numbers) and range. The function is called a *transfer (or activation) function* and usually returns values of the same type as its inputs and is usually constrained:

$$y = Neuron_W(X),$$

where W is the set of neuron parameters (including but not limiting to connection weights) and X is the set (or vector) of input connections.

The basic model of an artificial neuron, frequently referred to as a perceptron neuron, is presented in Fig. 3.1.

As can be seen, a simple artificial neuron realizes the following function (Haykin 1999; Bullinaria 2013; Aliev et al. 2004):

$$y = f\left(\theta + \sum_{i=1}^{n} w_i x_i\right),$$

where x_i is i-th input (connection) to the neuron, w_i is the weight of i-th (input) connection, $i = 1, \ldots n$, n is the number of inputs (input connections) to the neuron.

The function $f(.)$, whose argument is the total weighted input to the neuron (*NET*), is called an *activation function*. Activation function is usually a non-linear function for the neurons composing so-called hidden neurons (i.e. those being not classified as inputs or outputs) inside a network and can be linear for output

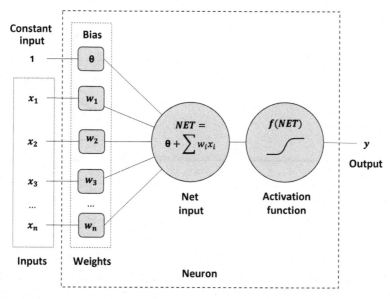

Fig. 3.1 The model of artificial neuron

neurons. The original choice of activation function for a perceptron was the step function (McCulloch-Pitts neurons):

$$f = \begin{cases} 1, \text{ if } NET \geq 0 \\ 0, \text{ otherwise} \end{cases}$$

Note that an activation function itself can have adjustable parameters. Widely used types of parameter-free activation functions are presented in Fig. 3.2.

Some activation functions with parameters are demonstrated in Fig. 3.3.

Some neuron models are defined just by the *transfer* function with specified set of neuron inputs and set of neuron parameters (see above for the general definition). This form is useful to describe particularly the network model utilized in Radial Basis Function (RBF) NNs. In this case, each connection weight can be described by a pair of real-valued parameters. The RBF neuron model realizes the transfer function such as:

$$y = \sqrt{\sum_i \left(\frac{x_i - c_i}{w_i} \right)^2},$$

where the parameters of the neuron are set of pairs of c_i and w_i for all input connections, $i = 1, \ldots, n$ or

Activation functions

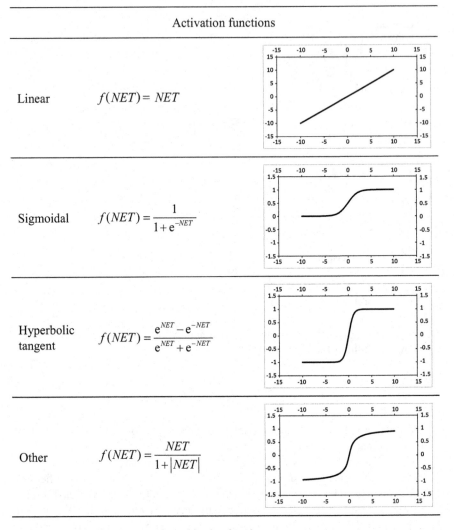

Linear	$f(NET) = NET$	
Sigmoidal	$f(NET) = \dfrac{1}{1 + e^{-NET}}$	
Hyperbolic tangent	$f(NET) = \dfrac{e^{NET} - e^{-NET}}{e^{NET} + e^{-NET}}$	
Other	$f(NET) = \dfrac{NET}{1 + |NET|}$	

Fig. 3.2 Examples of parameter-free activation functions

$$y = e^{-\left(\dfrac{\sum\limits_i \left(\dfrac{x_i - c_i}{w_i} \right)^2}{\sigma^2} \right)},$$

where the parameters in addition to sets of c_i and w_i include also σ.

Gaussian

$$f(NET) = \frac{1}{\sqrt{2\pi}\sigma}e^{-\frac{1}{2}\left(\frac{NET-\mu}{\sigma}\right)^2}$$

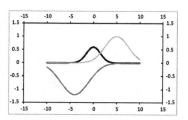

(parameters: μ and σ)

Parameterized sigmoidal

$$f(NET) = v + \frac{1}{1+e^{-\omega NET + \lambda}}$$

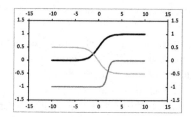

(parameters: v, ω, and λ)

Other

(parameters:)

Fig. 3.3 Examples of parameterized activation functions

3.1.2   Neural Network Architectures

Neural networks can be constructed of neurons using different architectures: lay-ered, feed-forward, recurrent etc. (Aliev et al. 2004; Bullinaria 2013).

Although it is quite possible that neurons in an artificial neural network are not organized in layers (e.g. fully-connected neural networks in which all neurons are interconnected), such cases are rare and we in this book will consider only layered network architectures.

In layered neural networks the neurons are organized into groups (layers) and the signal, after reception and processing by all neurons of a layer, is submitted to the succeeding layer and the process continues until the output layer is reached. Neurons of the same layer works in parallel. The layers between the input and output layers are called hidden layers. Number of hidden layers can be from 0 to infinity, in general.

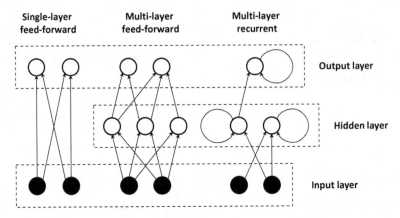

Fig. 3.4 Examples of network architectures

The input layer is usually just an input signal receiver and distributor. No other processing and no adjustable parameters (weights) are associated with units in this layer, although they are also called neurons – input neurons.

Layered neural networks can be single-layer and multi-layer. In single layer networks there is only one layer of processing neurons (no hidden layers), which is also an output layer. Multi-layer neural networks have two or more layers of processing neurons.

Neurons within the same layer do not communicate with each other in feed-forward networks and can have connections between themselves in recurrent neural networks. Different architectures (including mixtures of the above presented) can be used to serve as components of a neural network based systems (neural systems).

Figure 3.4 illustrates basic architectures of layered neural networks (Bullinaria 2013; Negnevitsky 2005).

3.1.3   Layered Neural Networks as Universal Approximators

As it was proven by many researchers, multi-layer neural networks can serve as universal approximators (Huang and Wu 2009). Let's consider the universal approximation theorem that can be stated as follows:

Theorem 3.1 *Universal approximation theorem (static case)* (Bullinaria 2013). Let $\varphi(.)$ be a non-constant, bounded, and monotone-increasing continuous function. Then for any continuous function $f(x_1, \ldots, x_m)$, where $x_i \in [0, 1]$, $i = 1, \ldots, m$ and $\varepsilon > 0$, there exists an integer M and constants u_j, θ_j, w_{kj}; $j = 1, \ldots, M$; $k = 1, \ldots, m$, such that

$$F(x_1, \ldots, x_m) = \sum_{j=1}^{M} u_j \varphi \left(\sum_{k=1}^{m} w_{jk} x_k - \theta_j \right)$$

is an approximation of $f(.)$, i.e.

$$|F(x_1, \ldots, x_m) - f(x_1, \ldots, x_m)| < \varepsilon$$

for all x_i, $i = 1, \ldots, m$, from the input space.

The above theorem applies to a feed-forward perceptron neural network with one hidden layer of sigmoidal activation function and the output layer of neurons with linear activation functions. In this case w_{jk} and θ_j are connection weights and thresholds of neurons in the hidden layer and u_j are connection weights of the output linear neuron. Therefore, it follows, that a feed-forward neural network (FFNN) with one hidden layer can approximate any continuous function, given the network enough hidden units.

An appropriate theorem exists also for the case of a dynamic system.

Theorem 3.2 *Universal approximation theorem (dynamic case)* (Bullinaria 2013). Any non-linear dynamical system can be approximated to any degree of accuracy by a recurrent neural network, with no restriction on the compactness of the state space, provided the network is given sufficient sigmoidal hidden neurons.

In a further section of the book we will consider approximation theorems proving capabilities of more complex neural networks (in particular, for logical neural networks with type-2 fuzzy neurons).

3.1.4   Design of Neural Networks

The class for a neural network, its architecture, and data type of signals is chosen based on the application problem solved by the network. For example, if few input–output data examples are available, there are expert based high-level data using which the input to output mapping can be defined by several rules or short table, or the model's logical transparency is important, then usually neuro-fuzzy systems are likely to be chosen. Otherwise, if there are large low-level input output data sets available, no expert knowledge, and a black-box model is acceptable, then a layered feed-forward network would be a better choice.

Complexity of the network is determined by many factors, including (Haykin 1999; Negnevitsky 2005; Aliev et al. 2004):

- Numbers of input and output neurons
- Number of hidden layers and neurons
- Neuron models
- Type of connections

- Activation functions

The complexity of the network and, especially, hidden layers and number of hidden neurons in them define the computational power of the network, i.e. complexity of model the network being able to simulate and limits of network's accuracy for its approximation. Accordingly, the optimal number of hidden layers and hidden units depends on many factors, among which are (Negnevitsky 2005; Bullinaria 2013):

- Complexity of the function or classification to be learned
- Numbers of input and output neurons
- Amount of noise in the training data
- Number and distribution of training data patterns
- Hidden neuron models, connections, and activation functions
- Training algorithm used

Best design should imply optimization of the network in terms of number of neurons, layers, and activation functions. The function or relation that the network needs to simulate is the main source of network complexity.

Unnecessary complexity is costly – not only it increases network operation and training time, it also impairs network generalization ability. A sensible strategy in choosing a best complexity is to do experiments trying a range of numbers of hidden layers and see which performs best.

Preprocessing, analysis, and filtering out wrong and noisy training data can also help to significantly simplify the architecture and improve the quality of layered networks. Numbers of hidden layers and neurons in logical networks may be controlled by the number and types of expert rules with the set of linguistic terms and membership functions. In latter case the optimization of number of rules can be done by clustering technique.

3.1.5   Training of Neural Networks

One of the most important features of neural networks is their ability to learn and generalize from a set of training patterns.

In general there are three types of learning (Bullinaria 2013; Negnevitsky 2005):

- Supervised learning (learning with a teacher – training);
- Reinforcement learning (with limited feedback) and
- Unsupervised learning (self-learning with no assistance)

This book will consider in some detail only the supervised learning algorithms for the most common types of neural networks, and hence, we will use the term *training*.

The aim of training is to train the network for the right (input–output) behavior. More specifically, the training is done to adjust the set of network parameters so that the network produces a desired output vector for any applied input vector. To this end usually a cost or error function is utilized which can evaluate the behavior of the network at any time (Aliev et al. 2011):

$$E(W) = \frac{1}{2} \sum_{p=1}^{P} \sum_{i=1}^{n_{output}} \left(y_{pi}^* - y_{pi} \right)^2 \qquad (3.1)$$

Here y_{pi}^* is the desired value (target) for output i for input value vector $\mathbf{x}_p$, y_{pi} is the actual output of the network for vector $\mathbf{x}_p$ ($y_{pi} = NN_W(\mathbf{x}_p)$), P is the number of input-to-output training patterns, and n_{output} is the number of outputs in the neural network.

The above function Eq. 3.1 "measures" how far the current network is from the desired (correctly trained) one. Note that the argument of the error function is the set of neural network neurons' parameters (e.g. connection weights) W including in rare cases some parameters defining network architecture.

Thus, the aim of neural network training is, given a set of examples (i.e. set of desired input-to-output patterns), to find parameter settings that minimize the error function $E(W)$. If there are N parameters, finding the parameter settings with minimum error involves searching through an N-dimensional Euclidean space.

In case of a neural network with fuzzy signals treated by crisp numerical optimization method, the fuzzy value of the error function is defuzzified by one of existing methods.

Training algorithms can be tightly linked to the overall model (input-to-output transfer function in analytic form) and architecture of a neural network. Then they are produced from customization of classical optimization techniques, such as various versions of coordinate descent and gradient descent algorithms. One of such algorithms is the well-known error back-propagation algorithm.

Alternatively, they can be independent of the model and architecture. Then they consider the neural network as a black-box computational system and use universal optimization techniques with no need for consideration of properties of the transfer function (e.g. derivatives etc.). The examples of the latter are evolutionary algorithms.

Usually the training starts with setting random initial weights and then adjusts them in small amounts until the required outputs are produced (Haykin 1999; Bullinaria 2013; Negnevitsky 2005):

$$w_{ij}(t+1) = w_{ij}(t) + \Delta w_{ij}(t)$$

Here $w_{ij}(t)$ stands for a network's parameter at time t, $w_{ij}(t+1)$ is the new parameter after change at time $(t+1)$ and $\Delta w_{ij}(t)$ is the value of change.

STEP 1.   Prepare the set of training patterns: $\left(\mathbf{x}_p, \mathbf{y}_p\right), p = 1, ..., P.$

STEP 2.   Create the network of the architecture $A\left[n_{input}, n_1, ..., n_{output}\right]$, defined required
           activation functions for hidden and output neurons.

STEP 3.   Generate randomly initial parameters W from the range $\left[-w_{max}, +w_{max}\right]$.

STEP 4.   Select an appropriate error function $E(W)$, e.g. $E(W) = \dfrac{1}{2} \sum\limits_{p=1}^{P} \sum\limits_{i=1}^{n_{output}} \left(y_{pi}^* - y_{pi}\right)^2$

STEP 5.   Compute the current neuron activations $y_i^{(l)}$ $(l = 1...L-1, i = 1, ...n_l)$ and the
           error function $E(W)$.

STEP 6.   If the value of $E(W)$ is acceptably small go to STEP 8.

STEP 7.   Apply the parameter update formula: $\Delta w_{ij}^{(l)} = -\eta \dfrac{\partial E(w_{ij}^{(l)})}{\partial w_{ij}^{(l)}}$ to each parameter
           $w_{ij}^{(l)}$ for each training pattern p.

STEP 8.   Go to STEP 5.

STEP 9.   Save the trained parameters.

Fig. 3.5 Gradient descent based neural network training algorithm

3.1.6   Gradient-Based Network Training Algorithms

As you know, gradient descent search involves repeated evaluation of the function
to be minimized – in this case the error – and its derivative (Haykin 1999; Bullinaria
2013; Negnevitsky 2005).

$$\Delta w_{ij} = -\eta \frac{\partial E\left(w_{ij}\right)}{\partial w_{ij}},$$

where η $(\eta > 0)$ is a coefficient called a learning (training) rate. Learning rate
specifies the step size in weight space to take for each iteration of the weight update
equation.

The complete scheme of the gradient descent training algorithm is presented in
Fig. 3.5.

Setting proper initial values for the parameters is important for the error back-
propagation algorithm because all parameters are treated the same way. If them all
set a same value, the changes will also done equally to all weights and the network
will never learn. A reasonable approach is to set them randomly in a range around
zero $[-w_{max}, +w_{max}]$, and set w_{max} a small value so as not to saturate the activation
function. The ranges of input and data in training set will also influence optimal

settings for the initial weights range. If the data is normalized in [0, 1], a good range for initial weights can be $[-0.001, +0.001]$.

Unfortunately, the optimal value for the learning rate η is very dependent on the training data and network architecture, so there are no reliable general prescriptions. The learning rate can change during the training process. As a starting learning rate values of 0.5 or 1.0 are often used. Smaller values for the rate lead to longer learning time – to get the minimum of the error function. On the other hand, larger values may make the error function to fluctuate or even diverge – weight changes will overshoot the error function minimum.

For a more specific case of gradient descent based neural network learning algorithm, let's consider a feed-forward perceptron neural network. Then, for a network without hidden layers ($L=2$), we will have (Negnevitsky 2005, Bullinaria):

$$\Delta w_{ij} = \eta \sum_p \left(y_i^* - y_i \right) f'(NET_i) \; x_j,$$

$$\Delta \theta_i = \eta \sum_p \left(y_i^* - y_i \right) f'(NET_i),$$

where $NET_i = \sum_k x_k w_{ik}$ and $f'(.)$ is the derivative of the network activation function.

For a multi-layer network, the update rule will change to:

$$\Delta w_{ij}^{(l)} = \eta \sum_p e_i^{(l)} f'\left(NET_i^{(l)}\right) \; y_j^{(l-1)},$$

$$\Delta \theta_i^{(l)} = \eta \sum_p e_i^{(l)} f'\left(NET_i^{(l)}\right),$$

Where:

$$NET_i^{(l)} = \sum_k y_k^{(l-1)} w_{ik}^{(l)}$$

$$e_i^{(L-1)} = y_i^* - y_i^{(L-1)}$$

$$e_i^{(l)} = \sum_k e_k^{(l+1)} w_{ki}^{(l+1)}, \quad l = 1, \ldots, L - 2$$

For simplicity in the above formulas index p is removed from the network activation and target identifiers ($x_{pi}^{(l)}$, $y_p^{(l)}$, y_p^*).

Clearly, having differentiable error functions and activation functions is crucial for the gradient descent algorithm to work. Different types of problem require different choices for these functions, and these may result in different weight update rules.

For the linear activation function:

$$f(NET) = NET$$
$$f'(NET) = 1.$$

For the sigmoidal activation function:

$$f(NET) = \frac{1}{1 + e^{-NET}}$$
$$f'(NET) = NET(1 - NET).$$

For the S-shaped activation function:

$$f(NET) = \frac{\widetilde{NET}}{1 + |\widetilde{NET}|}$$
$$f'(NET) = (1 - |NET|)^2.$$

Then, for example, if with the above given sum squared error (SSE) function we choose the sigmoidal activation function for the hidden layers and the linear activation function for the output layer, the formula for parameter change can be written as follows:

$$\Delta w_{ij}^{(L-1)} = \eta \sum_p e_i^{(L-1)} \, y_j^{(L-2)}$$

$$\Delta w_{ij}^{(l)} = \eta \sum_p e_i^{(l)} y_i^{(l)} \left(1 - y_i^{(l)}\right) \, y_j^{(l-1)}, \quad l = 1, \ldots, L - 2$$

3.2 Extension of Traditional Neural Networks by Introducing Fuzziness

The RBF and some other neuron models realizing parameterized convex and constrained in [0, 1] functions can be used to represent fuzzy variables in Fuzzy Neural Networks which we will consider in following sections. A neural model can be used to simulate fuzzy variables storing fuzzy terms of such traditional fuzzy models as triangle shaped (three parameters), trapezoidal (four parameters), and type-2 fuzzy (six and more parameters).

Fuzzy neuron models can be of two classes: (1) realizing data storage and processing components (e.g. membership functions of fuzzy and type-2 fuzzy terms) of a customized (unified and parallelized to resemble a neural network's architecture) version of fuzzy logical system, often referred to as a neuro-fuzzy

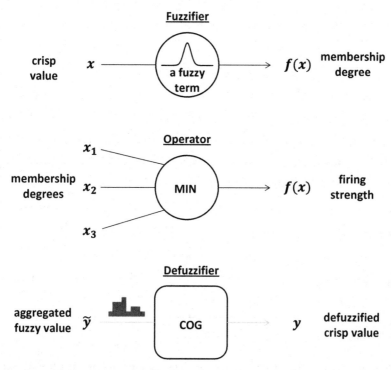

Fig. 3.6 Examples of fuzzy neurons

system and (2) being an extension of traditional artificial neurons to allow them accept, process, and output fuzzy and type-2 fuzzy values.

3.2.1 Fuzzy Logical Neuron Models (First Class)

A fuzzy neuron of the first class, i.e. a logical neuron usually realizes a parameterized membership function (a fuzzifier), a logical operator (AND, OR), or a type transformation operation (a defuzzifier, type-reducer, etc.) (Pedrycz 1993). Figure 3.6 illustrates some possible variants of fuzzy neurons.

The neurons implementing fuzzifiers are the units storing data of parameterized membership functions of type-1 fuzzy (ordinary fuzzy) and type 2 membership functions. The data-type of output data is real (crisp) numbers for type-1 fuzzy neuron models and fuzzy values for type-2 fuzzy neuron models. Very often for the sake of fast processing the output data-type for type-2 neurons is limited to the interval data-type.

Figure 3.7 illustrates examples of symmetric membership functions often used in fuzzy neurons realizing membership functions for type-1 fuzzy terms.

Bell-shaped (Generalized-bell) membership function

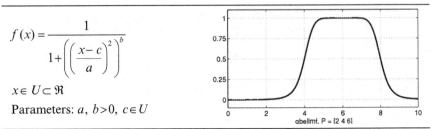

$$f(x) = \frac{1}{1 + \left(\left(\frac{x-c}{a} \right)^2 \right)^b}$$

$x \in U \subset \mathfrak{R}$

Parameters: a, $b > 0$, $c \in U$

Bell-shaped (Gaussian) membership function

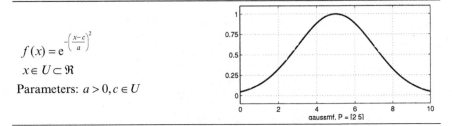

$$f(x) = e^{-\left(\frac{x-c}{a} \right)^2}$$

$x \in U \subset \mathfrak{R}$

Parameters: $a > 0, c \in U$

Fig. 3.7 Symmetric parameterized membership functions

When it is required that the neurons realize non-symmetric membership functions more complex neurons with more adjustable parameters can be used as well. Figure 3.8 demonstrates four parameter trapezoidal membership functions. For more efficient training, membership functions that are realized in neurons models are set to smooth functions. The continuity and differentiability of the functions realized in neurons is the requirement for the gradient based training methods. For evolutionary methods the differentiability is not important. However for efficient training of neurons using evolutionary algorithms, it is still important that the functions do not contain large flat regions like fuzzy core intervals in traditional trapezoidal membership functions. Smooth trapezoidal function presented in Fig. 3.8 used instead of traditional one can be more advantageous for efficient training.

More information on fuzzy logic oriented neural networks can be found in Sect. 3.4.

Various models exist also to represent type-2 fuzzy neurons. One of the simplified and quite useful models is the one that implements interval type-2 trapezoidal membership functions (Aliev et al. 2011; Abiyev et al. 2013; Castillo et al. 2013). The model simulates a type-2 fuzzy membership function (e.g. to represent a fuzzy type-2 term $\widetilde{A}$) by six real (crisp) numbers. The output of the single output neuron for an input real data-type value x is an interval data-type value μ_x.. This model will be considered in more detail in Sect. 3.6.

Trapezoidal membership function

$f(x) =$

$$\begin{cases} 0, & x \le a \vee x \ge d \\ \dfrac{x-a}{b-a}, & a < x < b \\ 1, & b \le x \le c \\ \dfrac{d-x}{d-c}, & c < x < d \end{cases}$$

$x \in U \subset \Re$

Parameters: $a \le b \le c \le d \in U$

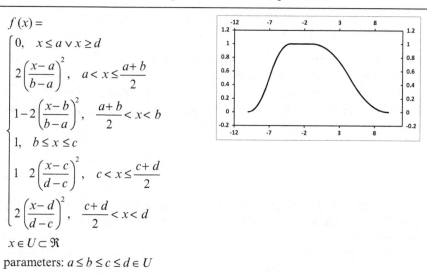

Smooth trapezoidal membership function

$f(x) =$

$$\begin{cases} 0, & x \le a \vee x \ge d \\ 2\left(\dfrac{x-a}{b-a}\right)^2, & a < x \le \dfrac{a+b}{2} \\ 1-2\left(\dfrac{x-b}{b-a}\right)^2, & \dfrac{a+b}{2} < x < b \\ 1, & b \le x \le c \\ 1\ 2\left(\dfrac{x-c}{d-c}\right)^2, & c < x \le \dfrac{c+d}{2} \\ 2\left(\dfrac{x-d}{d-c}\right)^2, & \dfrac{c+d}{2} < x < d \end{cases}$$

$x \in U \subset \Re$

parameters: $a \le b \le c \le d \in U$

Fig. 3.8 Trapezoidal membership functions for fuzzy neurons

3.2.2 Fuzzy Non-logical Neural Models (Second-Class)

A fuzzy neuron of the second class is much the same as an ordinary neuron except that it is able to get, process, and output fuzzy signals (Aliev et al. 2001, 2009). In other words, a fuzzy neuron implements a (crisp) function of one or more fuzzy arguments. Because the processing time is more critical in fuzzy case, for efficiency purposes some simplified logic of fuzzy mathematics can be used in fuzzy neurons.

As a variant of efficient activation function of a fuzzy (perceptron) neuron the following function can be used (Fig. 3.9):

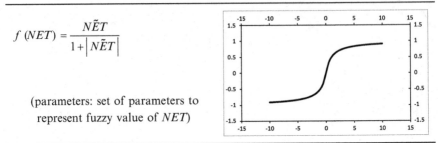

S-shaped (computation-efficient sigmoidal)

$$f\,(NET) = \frac{N\widetilde{E}T}{1 + \left| N\widetilde{E}T \right|}$$

(parameters: set of parameters to represent fuzzy value of NET)

Fig. 3.9 An example of activation function for a fuzzy neuron

$$f\left(N\widetilde{E}T\right) = \left.\frac{x}{1 + |x|}\right|x = N\widetilde{E}T \qquad (3.2)$$

Note that this function is differentiable and can be used even with gradient-based methods. Also this function is less computation-intensive than the sigmoidal function. If we use trapezoidal fuzzy numbers to represent fuzzy signals to and from the neuron, then, the net input signal to the neuron, $N\overline{E}T = \widetilde{T}(a, b, c, d)$, will generate the output signal:

$$f\left(N\widetilde{E}T\right) = \widetilde{T}\left(\frac{a}{1 + |a|}, \frac{b}{1 + |b|}, \frac{c}{1 + |c|}, \frac{d}{1 + |d|}\right)$$

It is correct to do so as the considered activation function is monotonously increasing. We cannot do it the same way for an activation function such as a Bell-shaped one.

Similar to fuzzy neurons there exist two classes of fuzzy (including type-2 fuzzy) layered neural networks: (1) fuzzy logic oriented inference (neuro-fuzzy) systems and (2) extension of traditional ANNs (e.g. fuzzy perceptron ANN).

Layered fuzzy and type-2 fuzzy neural networks of the first class are indeed rule based logical inference models put into a networked architecture. Indeed they are hybrids of fuzzy rule based inference system (or fuzzy logic, in narrow meaning) with connectionist system (i.e. artificial neural network). These networks consist of several (usually five and more) layers of neurons. Each layer includes neurons performing operations simulating those used in fuzzy and type-2 fuzzy logical inference schemes.

In a next section we will consider the mentioned-above neuro-fuzzy systems in more detail.

Fuzzy neural networks of the second class are an extension of traditional artificial neural networks. These fuzzy networks are based on a fuzzy model of neuron of the second class. A traditional model of neuron (e.g. a perceptron neuron with sigmoidal activation function) is extended to be able to receive, process, and send signals representing fuzzy (including type-2 fuzzy) numbers. The parameters of such network's neurons (e.g. the connection weights) model fuzzy numbers with one of fuzzy number representation models (e.g. triangular, trapezoidal, smooth trapezoidal etc.).

Of many different possible versions of fuzzy neural networks of the second class, in this book we will focus on the perceptron based feed-forward and recurrent fuzzy neural networks.

3.3 Fuzzy Feed-Forward and Recurrent Neural Networks

3.3.1 Basic Architecture and Operation of Fuzzy Feed-Forward Neural Networks

Feed-forward neural networks are simple but quite powerful and widely used type of artificial neural networks. A typical architecture of a feed-forward neural network of second class is demonstrated in Fig. 3.10 (no threshold connections are shown for simplicity) (Aliev et al. 2001).

To define architectures for traditional and fuzzy layered neural networks of second class (i.e. non neuro-fuzzy) across this book we will use the following notations.

$A[n_0, n_1, \ldots, n_{L-1}]$ will stand for the architecture of an L-layered (including the input layer 0) NN with $n_{input} \equiv n_0$ input (layer 0) neurons, following layer 1 with n_1 neurons, and so on, up to the final output layer $(L-1)$ with $n_{output} \equiv n_{L-1}$ neurons.

$x_j^{(l)}$ will denote j-th input signal to a neuron at layer l ($j = 1, \ldots, n_l$).

$y_i^{(l)}$ will denote output signal from i-th neuron at layer l ($i = 1, \ldots, n_l$).

Note that for an input neuron j ($j = 1, \ldots, n_{input}$) the following is true:

$$x_j \equiv x_j^{(0)} \equiv y_j^{(0)}.$$

For an output neuron i ($i = 1, \ldots, n_{output}$) the following is true: $y_i \equiv y_i^{(L-1)}$.

For a hidden neuron at layer l ($l = 1, \ldots L - 2$) the following is true: $x_j^{(l+1)} \equiv y_j^{(l)}$.

The vectors $\mathbf{x}^{(l)}$ and $\mathbf{y}^{(l)}$ are defined as follows:

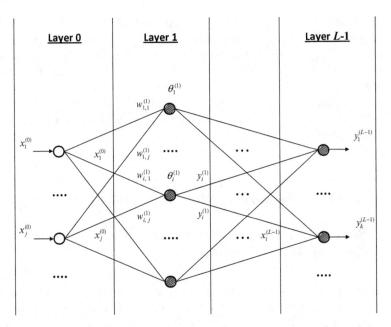

Fig. 3.10 Architecture of feed-forward neural network with fuzzy neurons of second class

$$\mathbf{x}^{(l)} \equiv \begin{pmatrix} x_1^{(l)} \\ x_2^{(l)} \\ \dots \\ x_{n_l}^{(l)} \\ 1 \end{pmatrix}$$

$$\mathbf{y}^{(l)} \equiv \begin{pmatrix} y_1^{(l)} \\ y_2^{(l)} \\ \dots \\ y_{n_l}^{(l)} \\ 1 \end{pmatrix}$$

W will denote the set of parameters including weights for all feed-forward connections and thresholds of the layered neural network.

Matrix $\mathbf{w}^{(l)}$ will denote weights connecting all neurons of layer l with all neurons of layer $(l+1)$. Thus for an L-layered NN set W will contain matrixes $\mathbf{w}^{(0)}, \mathbf{w}^{(1)}, \dots,$ $\mathbf{w}^{(L-1)}$. The individual parameters will be stored as follows:

Fig. 3.11 Network weights
and thresholds indexing

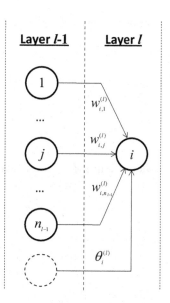

$$\mathbf{w}^{(l)} \equiv \begin{pmatrix} w_{1,1}^{(l)} & w_{1,2}^{(l)} & \cdots & w_{1,n_{l-1}}^{(l)} & \theta_1^{(l)} \\ w_{2,1}^{(l)} & w_{2,2}^{(l)} & \cdots & w_{2,n_{l-1}}^{(l)} & \theta_2^{(l)} \\ \cdots & \cdots & \cdots & & \cdots \\ w_{n_l,1}^{(l)} & w_{n_l,2}^{(l)} & \cdots & w_{n_l,n_{l-1}}^{(l)} & \theta_{n_l}^{(l)} \end{pmatrix}$$

Here $w_{i,j}^{(l)}$ is the weight of connection to neuron i at layer l from neuron j at the previous layer $(l-1)$, $\theta_i^{(l)}$ is the threshold parameter of neuron i at layer l (see Fig. 3.11).

Note that the total number of connection weights and thresholds (i.e. number of elements in the set W) for a feed-forward neural network (of the second class) is

$$N = (n_0 + 1)n_1 + (n_1 + 1)n_2 + \ldots + (n_{L-2} + 1)n_{L-1}.$$

Given particular values for the neural network parameters, and given values for the inputs, a neural network generates a value for each output:

$$y_i = NN_W(\mathbf{x}),$$

The operation of an L-layer feed-forward perceptron neural network at each layer l ($l = 1, \ldots, L-1$) can be described by the following equation:

$$y_i^{(l)} = f^{(l)}\left(\left(\sum w_{i,j}^{(l)} x_j^{(l)} \right) + \theta_i^{(l)} \right),$$

where $f^{(l)}(.)$ is the activation function used at network layer l.

In the vector form this can be written more compactly:

$$\mathbf{y}^{(l)} = f^{(l)} \left(\mathbf{w}^{(l)} \mathbf{x}^{(l)} \right)$$

Or, based on only the network activations as:

$$\mathbf{y}^{(l)} = f^{(l)} \left(\mathbf{w}^{(l)} \mathbf{y}^{(l-1)} \right)$$

For example, for a three-layer (input, one hidden, and output) NN, we will have the following input to output mapping:

$$\mathbf{y} = f^{(2)} \left(\mathbf{w}^{(2)} f^{(1)} \left(\mathbf{w}^{(1)} \mathbf{x} \right) \right),$$

where $\mathbf{x}$ and $\mathbf{y}$ are the network's input and output vectors, respectively.

3.3.2   Basic Architecture and Operation of Recurrent Neural Networks

A more complex type of layered neural networks is recurrent neural network. Architecture for a recurrent neural networks (of second class consisting of perceptron neurons) is shown in Fig. 3.12 (Aliev et al. 2008, 2009).

As can be seen, for a recurrent neural network (of second class) an additional set of parameters V is required to store the weights of feedback connections.

Matrix $\mathbf{v}^{(l)}$ denote all feed-back connection weights for all neurons at layer l. Thus for an L-layered recurrent NN (RNN) set V contains matrixes $\mathbf{v}^{(1)}, \mathbf{v}^{(2)}, \ldots, \mathbf{v}^{(L-1)}$:

$$\mathbf{v}^{(l)} \equiv \begin{pmatrix} v_{1,1}^{(l)} & v_{1,2}^{(l)} & \cdots & v_{1,n_l}^{(l)} \\ v_{2,1}^{(l)} & v_{2,2}^{(l)} & \cdots & v_{2,n_1}^{(l)} \\ \cdots & \cdots & \cdots & \cdots \\ v_{n_l,1}^{(l)} & v_{n_l,2}^{(l)} & \cdots & v_{n_l,n_l}^{(l)} \end{pmatrix}$$

In the considered version of recurrent neural network, at each layer except the input, there are feedback connections from outputs of all neurons (of that layer) to their inputs (at the same layer).

The total number of feed-back weights (i.e. number of elements in the set V) for a recurrent neural network (of the second class) is

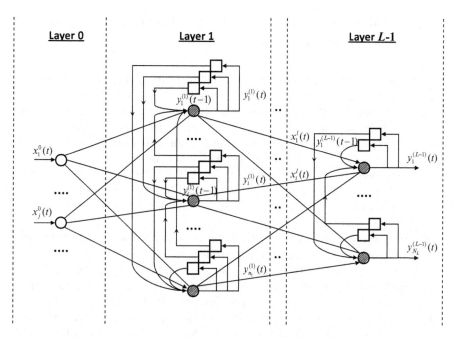

Fig. 3.12 Architecture of recurrent neural network with fuzzy neurons of second class

$$N = n_1^2 + n_2^2 + \ldots + n_{L-1}^2$$

The operation of an L-layer recurrent perceptron neural network at each layer l ($l = 1, \ldots, L-1$) can be described by the following equation:

$$y_i^{(l)}(t+1) = f^{(l)}\left(\left(\sum_{j=1, n_{l-1}} w_{i,j}^{(l)} x_j^{(l)}(t)\right) + \left(\sum_{k=1, n_l} v_{i,k}^{(l)} y_k^{(l)}(t)\right) + \theta_i^{(l)}\right).$$

In the vector form:

$$\mathbf{y}^{(l)}(t+1) = f^{(l)}\left(\mathbf{w}^{(l)}\mathbf{x}^{(l)}(t) + \mathbf{v}^{(l)}\mathbf{y}^{(l)}(t)\right)$$

or

$$\mathbf{y}^{(l)}(t+1) = f^{(l)}\left(\mathbf{w}^{(l)}\mathbf{y}^{(l-1)}(t) + \mathbf{v}^{(l)}\mathbf{y}^{(l)}(t)\right)$$

Note the input and output activations of a recurrent neural network are time-dependent.

3.3.3 Training of Neural Networks with Fuzzy Signals and Parameters

As with the case with the traditional neural networks the purpose is to develop an iterative procedure to determine network parameters (e.g. connection weights and biases) that would minimize a predefined error function. However, it is much more difficult to build the training procedure for fuzzy neural networks than for traditional and even for neuro-fuzzy systems because all the parameters (e.g. weights and biases) in a FNN are fuzzy numbers while they are crisp in a traditional or neuro-fuzzy systems.

Presently there are two approaches for training FNN. First approach is application of level-sets of fuzzy numbers and application of a gradient-based (e.g. the back-propagation) training algorithm.

The second approach to learning of regular and fuzzy NN involves evolutionary algorithms (EA) to minimize error function and determine fuzzy parameters (Aliev et al. 2001; Leung et al. 2003). In contrast to error back-propagation and other supervised learning algorithms, evolutionary algorithms are not based on information about the derivative of the activation function, and hence, they could potentially be more efficient for fuzzy feed-forward and especially for recurrent neural networks, computation of the derivative for which is an extremely difficult task involving processing of complex fuzzy Hessian or Jacobian matrices.

The application of the first approach has many difficulties among which are: extremely high computational burden, possible problems with regenerating valid fuzzy numbers from the level-sets, inability to deal with fuzzy numbers with non-differentiable membership functions etc.

Although there exist gradient-based algorithms in literature (Liu and Li 2004), the above mentioned problems as well as non-global optimizer nature of the such methods, especially, in presence of much more complex error functions, limit their application to real problems. For pure fuzzy neural networks (e.g. those requiring use of fuzzy arithmetic and fuzzy mathematical operations), application of the evolutionary methods are not only more natural (e.g. allowing more horizons for parallelism – one of key properties of NNs) but also more efficient, reliable, flexible, and even fast. That is why our further discussions on training algorithms for fuzzy networks will assume evolutionary based methods.

3.3.4 Evolutionary Algorithm Based Network Training Algorithms

As we mentioned above, gradient-based training methods can be faster than evolutionary algorithms based methods. However, they have noticeable deficiencies.

First, gradient-based based methods require knowledge of analytical network's computational model, differentiability of neuron activation functions and, in

Fig. 3.13 Multiple possible minima of the error function

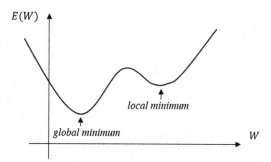

general, simpler models of neurons. For complex neuron models, like some fuzzy neurons with customized membership functions, generating a gradient-based learning method can lead to significant difficulties or can be even impossible (e.g. in case of non-differentiability of the model functions).

Secondly, the error function during the gradient-based training can quite easily end at the local minimum rather than the global minimum (Fig. 3.13) (Aliev et al 2001, 2009; Bullinaria 2013).

When applying an evolutionary training method, the error function is considered as a function producing its value using a computational model – in our case a neural network. The error function may be created using the predefined target set of input–output training pairs, like in the case of error back-propagation algorithm, or using a different logic. No information on the neural network model (including architecture, neuron models, and activation functions) is required. What is required is the set of optimization parameters – in our case all parameters W determining the network's architecture and operation (i.e. computation of neurons' activations).

In case of fuzzy and type-2 fuzzy connection weights, the list of parameters should consist of all crisp parameters required to describe the chosen fuzzy number format (e.g. four parameters to describe a single trapezoidal fuzzy weight).

For example, for an L-layered fuzzy perceptron recurrent neural network with symmetric triangle shaped fuzzy numbers used to represent fuzzy weights and thresholds, the total number N of parameters to be optimized by will be:

$$N = 2\big([(n_0 + 1)n_1 + (n_1 + 1)n_2 + \ldots + (n_{L-2} + 1)n_{L-1}] + [n_1^2 + n_2^2 + \ldots + n_{L-1}^2]\big)$$

Note that in the above formula, the multiplication by 2 is required because the symmetric triangular fuzzy number model needs two parameters to represent fuzzy numbers. For a more concrete three-layered recurrent network with two input, three hidden, and one output neurons (e.g. with architecture $A[2, 3, 1]$), the total number of parameters will be as many as 46! For a non-recurrent version with the same architecture the number of parameters will be just 26.

After obtaining a suitable error function $E(W)$, we create an evolutionary cost (or fitness function, which is more appropriate with a chosen evolutionary method):

$Cost(\mathbf{X})$, where $\mathbf{X}$ will be vector of parameters to optimize.

STEP 1.   Get access to the network output activation function $\mathbf{y} = NN(\mathbf{x})$, set of parameters W, error function $E(W)$, and, if required for the latter, prepare the set of training patterns: $(\mathbf{x}_p, \mathbf{y}_p)$, $p = 1, ..., P$.

STEP 2.   Compute number of evolutionary parameters to adjust. Prepare required encoding and decoding procedures for the parameters to be optimized.

STEP 3.   Initialize evolutionary process with required evolutionary parameters (size of population, crossover, mutation rates etc.) Select an appropriate cost function from the existing network error function $E(W)$, e.g.

$$Cost\ (\mathbf{X}) = Cost\ (Decode(W)) = E(W) = \frac{1}{2} \sum_{p=1}^{P} \sum_{i=1}^{n_{output}} \left(y_{pi}^* - NN(x_{pi}) \right)^2$$

STEP 4.   Continue minimization of function $Cost\ (\mathbf{X})$ using the evolutionary algorithm until the function falls below a predefined small value.

STEP 5.   Retrieve the optimal network parameters from the population: $W_{best} = Decode(\mathbf{X}_{best})$.

STEP 6.   Save the trained parameters.

Fig. 3.14 Evolutionary algorithm based neural network training algorithm

Note that the original parameters from W may be undergone some reversible transformations to convert them to evolutionary hypotheses (i.e. so-called evolutionary individuals – potential solutions) consider possible constraints (e.g. parameters of trapezoidal numbers $T(a, b, c, d)$, $a \leq b \leq c \leq d$), normalize (similar ranges for parameters), and format as required by method (e.g. bitstrings, chromosomes, etc.):

$$\mathbf{X} = Encode(W)$$

$$W = Decode(\mathbf{X}) = Encode^{-1}(\mathbf{X})$$

The general evolutionary algorithm based training for neural networks is illustrated in Fig. 3.14.

Note that the described evolutionary algorithm is suitable for all types of artificial networks: crisp, fuzzy, fuzzy-type-2, feed-forward, recurrent, logical etc.

3.4 Logic-Oriented Neural Networks for Fuzzy-Neuro Computing

3.4.1 Introductory Notes

In this section, we concentrate on the fundamentals and essential development issues of logic-driven constructs of fuzzy neural networks. These networks, referred to as logic-oriented neural networks, constitute an conceptual and computational framework that greatly benefits from the establishment of highly synergistic links between the technology of fuzzy sets (or granular computing, being more general) and neural networks. The most essential advantages of the proposed networks are twofold. First, the transparency of neural architectures becomes highly relevant when dealing with the mechanisms of efficient learning. Here the learning is augmented by the fact that domain knowledge could be easily incorporated in advance prior to any learning. This becomes possible given the compatibility between the architecture of the problem and the induced topology of the neural network. Second, once the training has been completed, the network can be easily interpreted and thus it directly translates into a series of truth- quantifiable logic expressions formed over a collection of information granules. The design process of the logic networks synergistically exploits the principles of information granulation, logic computing and underlying optimization including those biologically inspired techniques (such as particle swarm optimization, genetic algorithms and alike). We show how the logic blueprint of the networks is supported by the use of various constructs of fuzzy sets including logic operators, logic neurons, referential operators and fuzzy relational constructs.

Fuzzy sets and neural networks constitute two pillars of intelligent systems (Angelov 2004; Golden 1996; Jang et al. 1997; Kosko 1991; Mitra and Pal 1994, 1995; Nobuhara et al. 2005, 2006; Pal and Mitra 1999; Pedrycz et al. 1995; Pedrycz and Gomide 1998, 2007; Pedrycz 2004; Pedrycz and Reformat 2005; Setnes et al. 1998). Their research agendas are quite orthogonal and complementary to a significant extent. Fuzzy sets and granular computing, in general, (Pedrycz 2004) are aimed at representing knowledge in the form of information granules and forming highly interpretable relationships (models) at the granular level (Casillas et al. 2003; Dickerson and Lan 1995; Gobi and Pedrycz 2006). Neural networks are endowed with learning capabilities which make them highly adaptive. The distributed character of neural networks leads to evident interpretation difficulties; hence the networks are referred to as black box structures. Ideally, it would be beneficial to design hybrid architectures which come with the advantages of fuzzy systems and neural networks. Such constructs are referred to as logic- oriented neural networks where the name itself alludes to fuzzy set-based aspects of knowledge representation and a neuro-computing facet of learning capabilities. The crux of such logic- oriented networks is schematically visualized in Fig. 3.15. The networks of this form offer a unified framework in which neuro-computing and fundamentals of logic computing come hand in hand. In this sense, these networks combine the

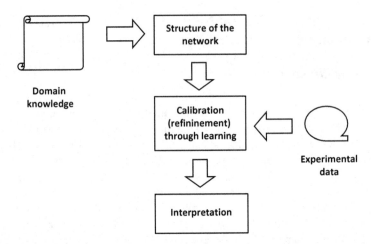

Fig. 3.15 Logic oriented networks

advantages of neural networks which are inherently associated with learning capabilities and benefits of logic architectures which manifest in their interpretability and transparency. The unified combination of learning flexibility and interpretability becomes a primordial feature of these networks. The transparency of the network greatly facilitates its development: we do not start from scratch and carry out intensive learning (which might be in some case quite inefficient and very much tedious) but rely on some prior knowledge which is instantaneously "downloaded" onto the structure of the network. The initial structure of the network is developed on a basis of some structural hints (points of navigation) conveyed by the existing prior knowledge. The experimental evidence (numeric data) is used to calibrate the network and further refine the structure. Once the learning has been completed the network can be interpreted as each neuron in its structure comes with a well-defined semantics. The aspect of interpretability of the networks requires attention in case of multiple input systems. The logic description in this case could be quite extended and therefore lead to some deterioration in terms of its interpretability given that the individual variables appear in the network. This shortcoming could be compensated by accepting a scenario that the network operates at the level of multivariable information granules (clusters) and in this way the number of inputs becomes equal to the number of information granules (which is typically far lower than the number of the original variables). Alternatively one could engage in some dimensionality reduction process prior to the design of the logic core of the network itself.

Given the environment of physical variables describing the surrounding world and an abstract view at the system under modeling, a very general view at the architecture of the fuzzy systems and logic-oriented neural networks can be portrayed as presented in Fig. 3.16 (Pedrycz and Gomide 2007; Pedrycz 2004).

It is worth distinguishing between three functional modules of the above architecture as each of them comes with well-defined objectives and roles. The input interface builds a collection of modalities (fuzzy sets and fuzzy relations) that are

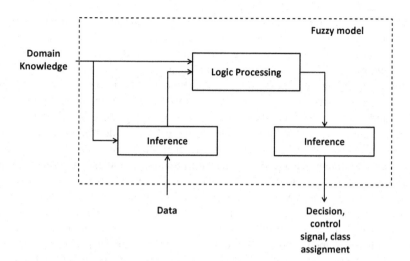

Fig. 3.16 A general view at the underlying architecture of fuzzy models

required to link the fuzzy model and its logic processing core with the external world. This processing core realizes all computing being carried out at the level of fuzzy sets (membership functions) already used in the interfaces. The output interface converts the results of granular (fuzzy) processing into the format acceptable by the modeling environment. In particular, this transformation may involve numeric values being the representatives of the fuzzy sets produced by the processing core. The interfaces could be present in different categories of the models yet they may show up to a significant extent. Their presence and relevance of the pertinent functionality depends upon the architecture of the specific fuzzy model and a way in which the model is utilized. The interfaces are also essential when the models are developed on a basis of available numeric experimental evidence as well as some prior knowledge provided by designers and experts.

3.4.2 Interfaces of Models

Interfaces form an essential functional entity of the networks. They are encountered in one way or the other in all fuzzy models. The interfaces are quite diversified in their formats and a ways in which input data become translated into the format acceptable at the level of the logic processing. In what follows, we concisely elaborate on the main categories of the interfaces, highlights their functionality, advantages and eventual limitations.

A single numeric variable can be granulated by using a small number of linguistic entities (say, small, medium, large, etc.) defined in a given universe of discourse. These linguistic terms come with a well-defined semantics that is reflective of the nature of the variable and a way in which such a variable is going to be used in the problem representation and perception as well as any further processing. In the literature there has been a substantial deal of studies devoted to the formation of a vocabulary of the linguistic terms and an analysis of the properties of the vocabulary itself. It has been noted that the main features of the associated fuzzy sets such as unimodality, coverage of the universe, distinguish-ability and alike are crucial to the retention of the meaning of the linguistic terms as sound semantic constructs. Furthermore we typically consider only a very limited number of linguistic terms where their number does not exceed 7 ± 2 terms. Different techniques of building such information granules are covered and contrasted in (Pedrycz 2004). More formally, consider "c" fuzzy sets $A_1, A_2, \ldots,$ A_c defined in some space $\mathbf{X}$. Any numeric input x_0 is quantified (perceived) in terms of the information granules by computing the degrees of membership $A_1(x_0),$ $A_2(x_0), \ldots, A_c(x_0)$. There could be different families of fuzzy sets realized in the form of Gaussian membership functions (in which case the properties highlighted above could be accomplished by a suitable distribution of the fuzzy sets and spreads of the fuzzy sets). Triangular fuzzy sets play also a visible role in the formation of the family of fuzzy sets.

From the computational standpoint, the original numeric entity positioned in the space of real numbers $\mathbf{R}$ results in a c-dimensional unit hypercube $[0, 1]^c$. It is worth stressing that each coordinate of the hypercube now comes with a logical flavor by being a degree of membership and functioning in the space of logic rather than physical components. From the computational standpoint, this type of mapping from $\mathbf{R}$ to $[0, 1]^c$ is a nonlinear mapping whose nonlinearity plays an essential role in further computing.

In a limit case, we can consider the use of a single membership function defined over the universe of discourse. Typically, to retain semantics of the linguistic term being used there, such membership function is monotonic, say being monotonically increasing.

The formation of fuzzy sets on individual variables leads to interpretable infor-mation granules, however, it comes at some cost which is an increase of dimen-sionality of the logic space in comparison with the original space. For instance, in case of "n" variables and the same number of information granules, we end up with $n * c$ variables to be used for further logic processing. When it comes to the formation of the rules, we may encounter a curse of dimensionality which becomes apparent in rule-based systems which involve such information granules. Consid-ering this, we can pursue a different avenue where all variables are granulated simultaneously. Fuzzy clustering comes as a viable alternative with this regard. Let us recall that information granules – fuzzy clusters are formed by minimizing a certain performance index (objective function). There is a large number of fuzzy clustering. One of the commonly used is a fuzzy C-means (FCM) which gives rise to a realization of "c" information granules in R^n whose membership grades are conveyed by a partition matrix U and a collection of "c" prototypes $v_1, v_2, \ldots, v_c$ defined in R^n. The computational details of this clustering technique is well-documented in the literature, cf. (Jang et al. 1997; Pedrycz and Gomide 2007;

Pedrycz 2004). It is worth noting that the information granules formed in this way are fully operational which means that any input x can be expressed in terms of the information granules. The underlying expression using which we compute the degrees of membership reads as follows:

$$u_{ik} = \frac{w_{ik}}{\sum_{j=1}^{c} \left(\frac{\|\mathbf{x_k} - \mathbf{v_i}\|}{\|\mathbf{x_k} - \mathbf{v_j}\|} \right)^{2/m-1}}$$

$$i, \ j = 1, 2, \ldots, c$$

where $m > 1$ is referred to as a fuzzification coefficient and $\| \, . \, \|$ denotes a certain distance function. Typically it is selected to be a Euclidean distance. In a nutshell, the granulation of information realized through fuzzy clustering produces "c" variables to be used at the level of logic processing. The dimensionality of this space could be either lower or higher than "n" depending upon the number of clusters. Quite often, though, we construct a relatively small number of information granules so for an even moderate dimensionality of the universe of discourse we have $c \ll n$. The dimensionality of the logic processing is reduced through information granulation which effectively improves further processing at the logic level.

The process of decoding deals with the transformation of the results of logic processing into the format acceptable by the modeling environment which is usually a certain numeric representative of the fuzzy sets (their levels of activation) translated into a single numeric quantity. There are several commonly encountered schemes including a mean of maxima and a center of gravity along with its variations. The quality of decoding depends on a number of design factors including the form of fuzzy sets and their number. With this regard, we have an interesting finding when dealing with triangular fuzzy sets with an 1/2 overlap between successive fuzzy sets, see Fig. 3.17 The essence of the scheme can be formulated in the form of the following proposition (Pedrycz and Gomide 2007).

Proposition Let us assume the following:

(a) The fuzzy sets of the codebook (termset) $\{A_i\}$, $i = 1, 2, \ldots, c$ form a fuzzy partition, $\sum_{i=1}^{c} A_i = 1$, and for each x in X at least one element of the codebook is activated, that is $A_i(x) > 0$.

(b) For each x only two neighboring elements of the codebook are "activated" that is $A_1(x) = 0, \ldots, A_{i-1}(x) = 0$, $A_i(x) > 0$, $A_{i+1}(x) > 0$, $A_{i+2}(x) = \ldots = A_c(x) = 0$.

(c) The decoding is realized as a weighted sum of the activation levels and the prototypes of the fuzzy sets v_i, namely $\hat{x} = \sum_{i=1}^{c} A_i(x) v_i$.

Then the elements of the codebook described by piecewise linear membership functions:

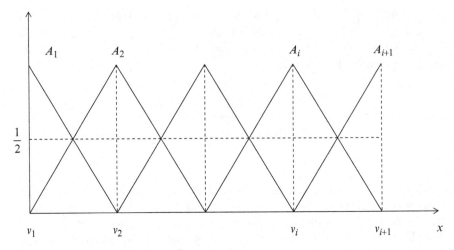

Fig. 3.17 An example of the code book composed of triangular fuzzy sets with an overlap of between each two neighboring elements of the codebook

$$A_i(x) = \begin{cases} \dfrac{x - v_{i-1}}{v_i - v_{i-1}} & \text{if } x \in [v_{i-1}, v_i] \\[2ex] \dfrac{x - v_{i+1}}{v_i - v_{i+1}} & \text{if } x \in [v_i, v_{i+1}] \end{cases}$$

give rise to the zero decoding error (lossless compression) meaning that $\hat{x} = x$.

Proof Consider any element x lying in the interval $[v_i, \ v_{i+1}]$ In virtue of (a) and (b) we can rewrite the decoding formula as follows:

$$\hat{x} = A_i(x)v_i + (1 - A_i(x))v_{i+1}$$

We request a lossless compression meaning that $\hat{x} = x$. In other words,

$$x = A_i(x)v_i + (1 - A_i(x))v_{i+1} \tag{3.3}$$

Then let us write down $A_i(x)$ (which describes the right-hand side of the membership function of A_i spread in-between v_i and v_{i+1}) by re-arranging the terms in Eq. 3.3. This leads to the expression

$$A_i(x)(v_i - v_i + 1) = x - v_{i+1}$$

and

$$A_i(x) = \frac{x - v_{i+1}}{v_i - v_{i+1}}$$

which reveals the piecewise linear character of the membership function of A_i. In the same fashion, we can deal with the left-hand side of the membership function of A_i by considering the interval $[v_{i-1}, v_i]$. In this case we can demonstrate that for all x in $[v_{i-1}, v_i]$ the membership comes in the form $A_i(x) = (x - v_{i-1})/(v_i - v_{i-1})$ which completes the proof.

3.4.3 Types of Fuzzy Neurons

The logic aspect of neuro-computing we intend to realize requires that the processing elements be endowed with the clearly delineated logic structure. We discuss several types of aggregative and referential neurons. Each of them comes with a clearly defined semantics of its underlying logic expression and is equipped with significant parametric flexibility necessary to facilitate substantial learning abilities.

Formally, these neurons realize a logic mapping from $[0, 1]^n$ to $[0, 1]$. Two main classes of the processing units exist in this category (Ciaramella et al. 2005, 2006; Hirota and Pedrycz 1994, 1999; Pedrycz 1991a, b, 1993; Pedrycz and Rocha 1993; Pedrycz 2007).

OR neuron realizes an *and* logic aggregation of inputs $\mathbf{x} = [x_1, x_2, \ldots, x_n]$ with the corresponding connections (weights) $\mathbf{w} = [w_1, w_2, \ldots, w_n]$ and then summarizes the partial results in an or-wise manner (hence the name of the neuron). The concise notation underlines this flow of computing, $y = OR(\mathbf{x}; \mathbf{w})$ while the realization of the logic operations gives rise to the expression (commonly referring to it as an s–t combination or $s - t$ aggregation).

$$y = \overset{n}{\underset{i=1}{S}} (x_i t w_i)$$

Bearing in mind the interpretation of the logic connectives (T-norms and T-conorms), the OR neuron realizes the following logic expression being viewed as an underlying logic description of the processing of the input signals

$$(x_1 \text{ and } w_1) \text{ or } (x_2 \text{ and } w_2) \text{ or } \ldots \text{ or } (x_n \text{ and } w_n)$$

Apparently the inputs are logically "weighted" by the values of the connections before producing the final result. In other words we can treat "y" as a truth value of the above statement where the truth values of the inputs are affected by the corresponding weights. Noticeably, lower values of w_i discount the impact of the corresponding inputs; higher values of the connections (especially those being positioned close to 1) do not affect the original truth values of the inputs resulting in the logic formula. In limit, if all connections $w_i, i = 1, 2, \ldots, n$ are set to 1 then the neuron produces a plain *or* -combination of the inputs, $y = x_1$ *or* x_2 *or*...*or* x_n. The values of the connections set to zero eliminate the corresponding inputs. Computationally, the OR neuron exhibits nonlinear characteristics (that is inherently implied by the use of the T-norms and T-conorms (that are evidently nonlinear mappings). The connections of the neuron contribute to its adaptive character; the changes in their values form the crux of the parametric learning.

AND neurons: The neurons in the category, denoted by $y = \text{AND}(\mathbf{x}; \mathbf{w})$ with $\mathbf{x}$ and $\mathbf{w}$ being defined as in case of the OR neuron, are governed by the expression.

$$y = \overset{n}{\underset{i=1}{T}} (x_i s w_i)$$

Here the *or* and *and* connectives are used in a reversed order: first the inputs are combined with the use of the T-conorm and the partial results produced in this way are aggregated *and*-wise.

Higher values of the connections reduce impact of the corresponding inputs. In limit $w_i = 1$ eliminates the relevance of x_i. With all w_i set to 0, the output of the AND neuron is just an *and* aggregation of the inputs.

$y = x_1$ and x_2 and $\ldots$ and x_n

The influence of the connections and the specific realization of the triangular norms on the mapping completed by the neuron. The connections of the neurons are set to 0.1 and 0.7 with intent to visualize their effect on the produced characteristics of the neuron.

Let us conclude that the neurons are highly nonlinear processing units whose nonlinear mapping depends upon the specific realizations of the logic connectives. They also come with potential plasticity whose usage becomes critical when learning the networks including such neurons.

At this point, it is worth contrasting these two categories of logic neurons with "standard" neurons we encounter in neuro-computing. The typical construct there comes in the form of the weighted sum of the inputs $x_1, x_2, \ldots, x_n$ with the corresponding connections (weights) $w_1, w_2, \ldots, w_n$ being followed by a nonlinear (usually monotonically increasing) function that reads as follows:

$$y = g\left(\mathbf{w}^{\mathsf{T}} x + \tau\right) = g\left(\sum_{i=1}^{n} w_i x_i + \tau\right)$$

where $\mathbf{w}$ is a vector of connections, τ is a constant term (bias) and g denotes some monotonically non-decreasing nonlinear mapping. The other less commonly encountered neuron is a so-called π-neuron. While there could be some variations as to the parametric details of this construct; we can envision the following realization of the neuron:

$$y = g\left(\prod |x_i - t_i|^{w_i}\right)$$

where $t = [t_1 \; t_2 \; \ldots \; t_n]$ denotes a vector of translations while $\mathbf{w}$ (>0) denotes a vector of all connections.

As before, the nonlinear function is denoted by "g". While some superficial and quite loose analogy between these processing units and logic neurons could be derived, one has to cognizant that these neurons do not come with any underlying logic fabric and hence cannot be immediately interpreted.

Let us make two additional observations about the architectural and functional facets of the logic neurons we have introduced so far.

Incorporation of the bias term (bias) in the fuzzy logic neurons: In analogy to the standard constructs of a generic neuron as presented above, we could also consider a bias term, denoted by $w_0 \in [0, \; 1]$ which enters the processing formula of the fuzzy neuron in the following manner:

For the OR neuron:

$$y = \mathop{S}_{i=1}^{n} (x_i t w_i) s w_0$$

For the AND neuron:

$$y = \mathop{T}_{i=1}^{n} (x_i s w_i) t w_0$$

We can offer some useful interpretation of the bias by treating it as some non-zero initial truth value associated with the logic expression of the neuron. For the OR neuron it means that the output does not reach values lower than the assumed threshold. For the AND neuron equipped with some bias, we conclude that its output cannot exceed the value assumed by the bias. The question whether the bias is essential in the construct of the logic neurons cannot be fully answered in advance. Instead, we may include it into the structure of the neuron and carry out learning. Once its value has been obtained, its relevance could be established considering the specific value it has been produced during the learning. It may well be that the optimized value of the bias is close to zero for the OR neuron or close to one in the case of the AND neuron which indicates that it could be eliminated without exhibiting any substantial impact on the performance of the neuron.

Dealing with inhibitory character of input information: Owing to the monotonicity of the *T-norms* and *T-conorms*, the computing realized by the neurons exhibits an excitatory character. This means that higher values of the inputs (x_i) contribute to the increase in the values of the output of the neuron. The inhibitory nature of computing realized by "standard" neurons by using negative values of the connections or the inputs is not available here as the truth values (membership grades) in fuzzy sets are confined to the unit interval. The inhibitory nature of processing can be accomplished by considering the complement of the original input, $\bar{x}_i = 1 - x_i$. Hence when the values of x_i increase, the associated values of the complement decrease and subsequently in this configuration we could effectively treat such an input as having an inhibitory nature.

The essence of referential computing deals with processing logic predicates. The two-argument (or generally multivariable) predicates such as *similar, included in, dominates* (Pedrycz and Rocha 1993) are essential components of any logic description of a system. In general, the truth value of the predicate is a degree of satisfaction of the expression $P(x, a)$ where "a" is a certain reference value (reference point). Depending upon the meaning of the predicate (P), the expression $P(x, a)$ reads as "x is similar to a", "x is included in a", "x dominates a", etc. In case of many variables, the compound predicate comes in the form $P(x_1, x_2, \ldots, x_n, a_1, a_2, \ldots, a_n)$ or more concisely $P(\mathbf{x}; \mathbf{a})$ where $\mathbf{x}$ and $\mathbf{a}$ are vectors in the n-dimensional unit hypercube. We envision the following realization of $P(\mathbf{x}; \mathbf{a})$:

$$P(\mathbf{x}; \mathbf{a}) = P(x_1, a_1) \text{ and } P(x_2, a_2) \text{ and } \ldots \text{ and } P(x_n, a_n) \qquad (3.4)$$

meaning that the satisfaction of the multivariable predicate relies on the satisfaction realized for each variable separately. As the variables could come with different level of relevance as to the overall satisfaction of the predicates, we represent this effect by some weights (connections) $w_1, w_2, \ldots, w_n$ so that Eq. 3.4 can be expressed in the following form:

$$P(\mathbf{x}; \mathbf{a}, \mathbf{w}) = [P(x_1, a_1) \text{ or } w_1] \text{ and } [P(x_2, a_2) \text{ or } w_2] \text{ and } \ldots \text{ and } [P(x_n, a_n) \text{ or } w_n]$$

Taking another look at the above expression and using a notation $z_i = P(x_i, a_i)$, it corresponds to a certain AND neuron $y = AND(\mathbf{z}; \mathbf{w})$ with the vector of inputs $\mathbf{z}$ being the result of the referential computations done for the logic predicate. Then the general notation to be used reads as $REF(\mathbf{x}; \mathbf{w}; \mathbf{a})$. In the notation below, we explicitly articulate the role of the connections

$$y = \overset{n}{\underset{i=1}{T}} (REF(x_i, a_i)sw_i)$$

In essence, as visualized in Fig. 3.18, we may conclude that the reference neuron is a realized as a two-stage construct where first we determine the truth values of the predicate (with $\mathbf{a}$ being treated as a reference point) and then treat these results as the inputs to the AND neuron.

So far we have used the general term of predicate-based computing not confining ourselves to any specific nature of the predicate itself. Among a number of available possibilities of such predicates, we discuss the three of them, which tend to occupy an important place in logic processing. Those are inclusion, dominance and match (similarity) predicates. As the names stipulate, the predicates return truth values of satisfaction of the relationship of inclusion, dominance and similarity of a certain argument "x" with respect to the given reference "a". The essence of all these calculations is in the determination of the given truth values and this is done in the

Fig. 3.18 A schematic
view of computing realized
by a reference neuron an
involving two processing
phases (referential
computing and aggregation)

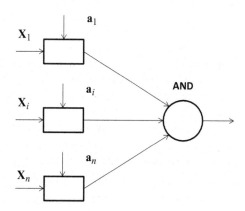

carefully developed logic framework so that the operations retain their semantics
and interpretability. What makes our discussion coherent is the fact that the
proposed operations originate from triangular norms. The inclusion operation, as
discussed earlier, denoted by $\subset$, is modeled by an implication $\Rightarrow$ that is induced by
a certain left continuous T-norm (Pedrycz and Gomide 1998, 2007)

$$a \Rightarrow b = \sup\{c \in [0, 1] | a t c \leq b\}, a, b \in [0, 1]$$

For instance, for the product the inclusion takes on the form $a \Rightarrow b = \min(1, b/a)$.
The intuitive form of this predicate is self-evident: the statement "x is included in a"
and modeled as $INCL(x, a) = x \Rightarrow a$ a comes with the truth value equal to 1 if x is
less or equal to a (which in other words means that x is included in a) and produces
lower truth values once x starts exceeding the truth values of "a". Higher values of
"x" (those above the values of the reference point "a") start generating lower truth
values of the predicate. The dominance predicate acts in a dual manner when
compared with the predicate of inclusion. It returns 1 once "x" dominates "a"
(so that its values exceeds "a") and values below 1 for x lower than the given
threshold. The formal model can be realized as $DOM(x, a) = a \Rightarrow x$. With regard to
the reference neuron, the notation is equivalent to the one being used in the previous
case, that is $DOM(\mathbf{x}; \mathbf{w}; \mathbf{a})$ with the same meaning of $\mathbf{a}$ and $\mathbf{w}$.

The similarity (match) operation is an aggregate of these two, $SIM(x, a) = INCL$
(x, a) t $DOM(x, a)$ which is appealing from the intuitive standpoint: we say that x is
similar to a if x is included in a and x dominates a. Noticeably, if $x = a$ the predicate
returns 1; if x moves apart from "a" the truth value of the predicate becomes
reduced. The resulting similarity neuron is denoted by $SIM(\mathbf{x};\mathbf{w},\mathbf{a})$ and reads as

$$y = \overset{n}{\underset{i=1}{T}} (SIM(x_i, a_i) s w_i)$$

The reference operations form an interesting generalization of the threshold
operations. Consider that we are viewing "x" as a temporal signal (that changes

over time) whose behavior needs to be monitored with respect to some bounds (α and β). If the signal does not exceed some predefined threshold α then the acceptance signal should go off. Likewise we require another acceptance mechanism that indicates a situation where the signal does not go below another threshold value of β. In the case of fuzzy predicates, the level of acceptance assumes values in the unit interval rather than being a Boolean variable. Furthermore the strength of acceptance reflects how much the signal adheres to the assumed thresholds.

It is worth noting that by moving the reference point to the origin or the 1-vertex of the unit hypercube (with all its coordinates being set up to 1), the referential neuron starts resembling the aggregative neuron. In particular, we have:

- for $\mathbf{a} = \mathbf{1} = [1\ 1\ 1\ \ldots\ 1]$ the inclusion neuron reduces to the AND neuron and
- for $\mathbf{a} = \mathbf{0} = [0\ 0\ 0\ \ldots\ 0]$ the dominance neuron reduces to the AND neuron.

One can draw a loose analogy between some types of the referential neurons and the two categories of processing units encountered in neuro-computing. The analogy is based upon the *local* versus *global* character of processing realized therein. Perceptrons come with the global character of processing. Radial basis functions realize a local character of processing as focused on receptive fields. In the same vein, the inclusion and dominance neurons are after the global nature of processing while the similarity neuron carries more confined and local processing.

There is an interesting taxonomy of referential neurons. We distinguish between two categories of these processing elements. The first group is formed by homogeneous neurons, i.e. those which encounter the same type of logic predicate (P) used in the underlying processing. Refer also to (*xx*). As shown above the underlying logic expression reads as

$$y = (P(x_1, r_1)w_1) \quad and \quad (P(x_2, r_2)w_2) \quad and \quad \ldots \quad and(P(x_n, r_n)w_n)$$

where r_i and w_i are the point of reference and the corresponding weight associated with the i-th variable. Thus all inputs are processed making use of the same predicate – could that be inclusion, dominance, tolerance, similarity etc.

In the second groups of referential neurons we admit a higher level of diversity by allowing different predicates P_i associated with the individual inputs. The general logic expression comes in the following format:

$$y = (P_1(x_1, r_1)w_1) \quad and \quad (P_2(x_2, r_2)w_2) \quad and \quad \ldots \quad and(P_n(x_n, r_n)w_n)$$

Referential computing is reflective of processing logic constraints conveyed by logic predicates when dealing with existing domain knowledge and encapsulating it in the structural format of the network or its part.

For instance, we can represent existing constraints for the variables $x_1, x_2, \ldots, x_n$ which are spelled out as

x_1 should not exceed g_1 and x_2 should not exceed g_2 and $\ldots$
x_n should not exceed g_n

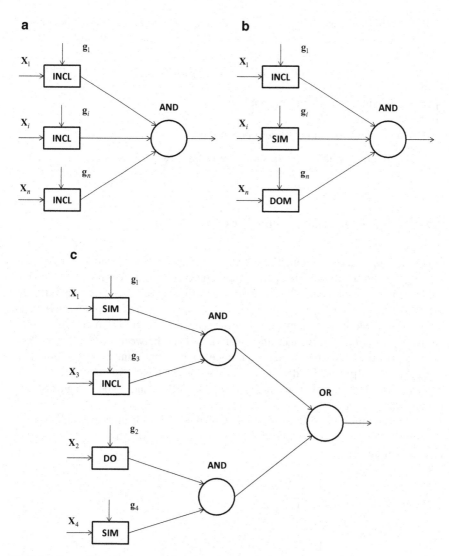

Fig. 3.19 Examples of logic expressions involving predicates and their realization in the form of logic neurons and architectures of logic-oriented networks

in the form of a single inclusion neuron, Fig. 3.19a where the weights are used to calibrate the relevance (importance) of the corresponding constraint. Note that both x_i and g_i standing in the above compound predicate assume values in the unit interval (so they could the result of the *transformation(xx)*). One might capture a variety of constraints.

x_1 should not exceed g_1 and x_2 should be similar to g_2 and ...
x_n should dominate g_n.

By considering various predicates; see Fig. 3.19b. Interestingly, one could consider a variety of logic constraints of different dimensionality that is each referential neuron might have different number of inputs depending on the nature of locally identified constraints. For instance, given three inputs x_1, x_2, and x_3, we have

x_1 should be similar to g_1 and x_3 should not exceed g_3

or

x_2 should dominate g_2 and x_3 should be similar to g_4.

This logic description translates into the logic network illustrated in Fig. 3.19c.

3.4.4 Uninorms and Unineurons

Uninorms constitute an interesting generalization of T-norms and T-conorms by binding the two standard logic operators encountered in logic and fuzzy sets (Yager and Rybalov 1996; Yager 2001). More formally, a uninorm is a mapping u: $[0, 1]^2 \rightarrow [0, 1]$ that satisfies the properties of *commutativity*, *monotonicity* and *associativity*. More importantly, by introducing the identity element "g" which varies between 0 and 1, we can implement switching between the "and" and "or" properties of the logic operators. For instance, given input x and identity element g, $u(x, g) = x$, $g \in [0, 1]$. Evidently when $g = 0$ we end up with the "or" type of aggregation, $u(x, 0) = x$ and when $g = 1$, it returns the "and" type of aggregation, namely $u(x, 1) = x$.

In the literature, there are many types of realization of uninorms (Yager and Rybalov 1996). Here, we choose the following equation which turns to be flexible and easy to interpret:

$$u(x, y, g) = \begin{cases} gt\left(\dfrac{x}{g}, \dfrac{y}{g}\right) & x, y \in [0,\ g] \\[2mm] g + (1 - g)s\left(\dfrac{x - g}{1 - g}, \dfrac{y - g}{1 - g}\right) & x, y \in [g,\ 1] \\[2mm] \min(x, y)\ \ or\ \ \max(x, y) & \text{otherwise} \end{cases}$$

In the above equation, x and y are two inputs between 0 and 1, "t" denotes a certain *T-norm*; "s" stands for some *T-conorm* (*S-norm*).

Incorporating uninorms into fuzzy neurons discussed in the previous section subsequently constructs them as unineurons. Unineurons, like general neurons, are treated as n-input nonlinear static processing units that map elements in the unit hypercube $[0, 1]^n$ into elements in the unit interval of $[0, 1]$. There are two levels of the logic processing carried out in the processing units. More specifically, given a collection of inputs $\mathbf{x} = [x_1\ x_2\ \ldots\ x_n]$ and the parameters of unineurons including

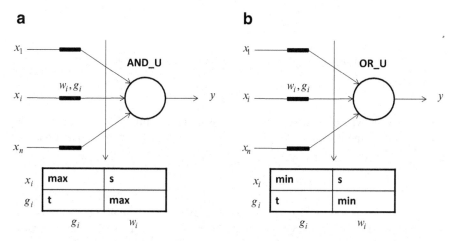

Fig. 3.20 A schematic view at logic processing realized by unineurons: (**a**) AND unineuron and (**b**) OR unineuron

connections $\mathbf{w} = [w_1 \ w_2 \ \ldots \ w_n]$ and neutral points $\mathbf{g} = [g_1 \ g_2 \ \ldots \ g_n]$, at the first level we exploit the use of uninorms by combining individual input x_i with corresponding connection w_i giving rise to the expression $u(x_i; \ w_i, g_i)$ with g_i being the neutral point of this uninorm, and the resulting aggregation is realized at the second level.

Two fundamental categories of logic neurons are introduced here, which will be referred to as AND and OR unineurons, or AND_U and OR_U when using shorthand notation.

As shown in Fig. 3.20a, given a collection of "n" inputs $\mathbf{x} = [x_1 \ x_2 \ \ldots \ x_n]$ and the parameters of unineurons including connections $\mathbf{w} = [w_1 \ w_2 \ \ldots \ w_n]$ and neutral points $\mathbf{g} = [g_1 \ g_2 \ \ldots \ g_n]$, the AND_U processes them in the following format:

$$y = AND\_U(\mathbf{x}; \mathbf{w}, \mathbf{g})$$

or in the coordinatewise form, it can be rewritten as

$$y = \overset{n}{\underset{i=1}{T}} \left(u \ (x_i; \ w_j g_i) \right)$$

In the above equation, the uninorm operation is realized in the form governed by some uninorm. As becomes quite apparent the name of the unineuron is implied by the *and* type of aggregation of the individual inputs. Moreover, the standard AND neurons are subsumed by the AND_U neurons when using the zero vector of the neutral points $\mathbf{g} = \mathbf{0}$, that is

$$y = \mathop{T}_{i=1}^{n} \left(u\left(x_i; w_j, 0\right) \right) = \mathop{T}_{i=1}^{n} \left(s\left(x_i; w_j\right) \right)$$

Similarly, an n-input single output g realized by this processing unit

$$y = OR\_U(\mathbf{x}; \mathbf{w}, \mathbf{g})$$

concerns an or-type of aggregation of the partial results produced by the uninorm combination of the corresponding inputs. We can rewrite the above equation as

$$y = \mathop{S}_{i=1}^{n} \left(u\left(x_i; w_j, g_i\right) \right)$$

Where $S(s)$ stands for any S-norm (T-conorm). Also, the standard OR neurons are subsumed by the OR_U neurons when all the entries of the neutral points $\mathbf{g}$ are all ones($\mathbf{g} = 1$), that is

$$y = \mathop{S}_{i=1}^{n} \left(u\left(x_i; w_j, 1\right) \right) = \mathop{S}_{i=1}^{n} \left(t(x_i; wj) \right)$$

3.4.5   Architectures of Fuzzy Logic-Oriented Neural Networks

Logic neurons give rise to a surprisingly rich and diversified landscape of neural architectures. They are reflective of the existing domain knowledge which can be articulated in terms of logic predicates and structured into compound statements. We discuss several representative categories of such neural networks starting with a regular topology of logic processors and presenting ideas of highly regular architectures which are typical for such constructs as fuzzy cognitive maps.

The typical logic network that is at the center of logic processing originates from the two-valued logic and comes in the form of the famous Shannon theorem of decomposition of Boolean functions. Let us recall that any Boolean function $\{0, 1\}^n \rightarrow \{0, 1\}$ can be represented as a logic sum of its corresponding miniterms or a logic product of max-terms. By a min-term of "n" logic variables x_1, x_2 ,..., x_n we mean a logic product involving all these variables either in direct or complemented form. Having "n" variables we end up with 2^n min-terms starting from the one involving all complemented variables and ending up at the logic product with all direct variables. Likewise by a max-term we mean a logic sum of all variables or their complements. Now in virtue of the decomposition theorem, we note that the first representation scheme involves a two-layer network where the first layer consists of AND gates whose outputs are combined in a single OR gate. The converse topology occurs for the second decomposition mode: there is a single layer of OR gates followed by a single AND gate aggregating or-wise all partial results.

Fig. 3.21 A topology of the
logic processor in its AND–
OR mode

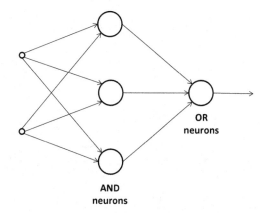

OR
neurons

AND
neurons

The proposed network (referred here as a logic processor) generalizes this
concept as shown in Fig. 3.21. The OR-AND mode of the logic processor comes
with the two types of aggregative neurons being swapped between the layers. Here
the first (hidden) layer is composed of the OR neuron and is followed by the output
realized by means of the AND neuron.

The logic neurons generalize digital gates. The design of the network (viz. any
fuzzy function) is realized through learning. If we confine ourselves to $\{0, 1\}$ values,
the network's learning becomes an alternative to a standard digital design, especially a minimization of logic functions. The logic processor translates into a
compound logic statement (we skip the connections of the neurons to underline
the underlying logic content of the statement):

$$\text{if} \left(input_i \, and \, \dots \, and \, input_j\right) \, or \, \left(input_d \, and \, \dots \, and \, input_f\right) \text{ then } output$$

The logic processor's topology (and underlying interpretation) is standard. Two
LPs can vary in terms of the number of AND neurons, their connections but the
format of the resulting logic expression is quite uniform (as a sum of generalized
min-terms).

The architectures of fuzzy neural networks discussed so far are concerned with
some static mappings between unit hypercubes, $[0, 1]^n - [0, 1]^m$. The dynamics of
systems modeled in a logic manner can be captured by introducing some feedback
loops into the structures. Some examples of the networks with the feedback loops
are illustrated in Fig. 3.22. They are based on the logic processor we have already
studied. In the first case, the feedback loop is built between the output layer and the
inputs of the AND neurons, Fig. 3.22a. The form of the feedback (that could be
either positive – excitatory or negative – inhibitory) is realized by taking the
original signal (thus forming an excitatory loop-higher values of input "excite"
the corresponding neuron) or its complement (in which case we end up with
the inhibitory effect – higher values of the signal suppress the output of the
corresponding neuron). The strength of the feedback loop is modeled by the

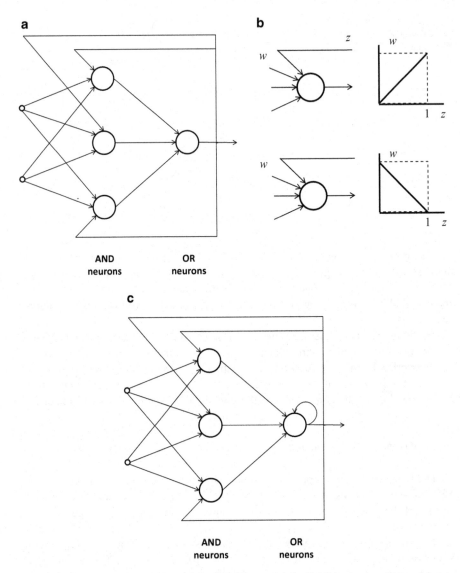

Fig. 3.22 Fuzzy neural networks with feedback loops: (**a**) feedback between output and input layer, (**b**) notation used in describing excitatory and inhibitory feedback mechanisms along with the underlying computing, and (**c**) feedback loop between output layer and the input layer as well as itself

numeric of the connection. To visualize the effect of inhibition or excitation, we use the dot notation as illustrated in Fig. 3.22b. The small dot corresponds to the complement of the original variable. Another example of the feedback loops shown in Fig. 3.22c involves the two layers of the neurons of the network.

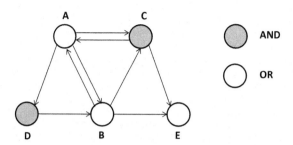

Fig. 3.23 An example of a fuzzy cognitive map whose nodes are realized in the form of some logic expressions (AND and OR neurons). The inhibitory effect is realized by taking the complement of the activation level of the interacting node (here indicated symbolically by a *small dot*)

The semantics of such networks is straightforward. Any prior knowledge is essential in the formation of the structure of the network itself. Then the problem can be directly mapped onto the network.

The structure of the network offers a great deal of flexibility and is far less rigid than the fuzzy neural networks where typically the nodes (neurons) are organized into some layers. An example comes under the rubric of fuzzy cognitive maps (Kosko 1991) which can be represented in the form of logic neurons. The underlying idea is that the individual nodes of the fuzzy cognitive maps could be realized as some logic expressions and implemented as AND or OR neurons. The connections could assume values in [0, 1] and the inhibitory effect can be realized by taking the complement of the activation level of the node linked to the one under consideration. An example of the logic-based fuzzy cognitive map is illustrated in Fig. 3.23.

3.4.6 Interpretation of the Fuzzy Neural Networks

Each logic neuron comes with a well-defined semantics that is directly associated with the underlying logic. OR neurons realize a weighted or combination of their inputs. The higher are the value of the connection, the more essential becomes the corresponding input. For the AND neuron the converse relationship holds: lower values of the connections imply higher relevance of the corresponding inputs. While these two arrangements are taken into consideration, we can generate a series of rules generated from the network. We start with the highest value of the OR connection and then translate the corresponding AND neuron into the *and* combination of the inputs. Again the respective inputs are organized according to their relevance proceeding with the lowest value of the corresponding connection.

For illustrative purpose, we consider the fuzzy neural network shown in Fig. 3.24. The rules can be immediately enumerated from this structure. In addition, we can order them by listing the most dominant rules first. This ordering refers to the relevance of the rules and the importance of each condition standing there. We

Fig. 3.24 An example
fuzzy neural network

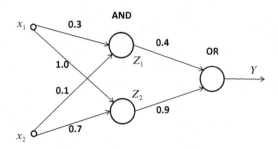

start from the output (OR) node and enumerate the inputs in the order implied by the values of the connections of the neuron. We get

$$\text{IF } z_{2|0.9} \text{ or } z_{1|104} \text{ THEN } y$$

where the subscripts (0.9 and 0.4, respectively) denote the confidence level we associate with the corresponding inputs of the neuron.

Next we expand the terms z_1 and z_2 using the original inputs x_1 and x_2; in this case we report the inputs in the increasing order of the connections starting from the lowest one. Recall that the connections of the AND neuron realize a masking effect; the higher the value of the connection, the stronger the masking effect. In particular, if the connection is equal to 1, we can drop the associated input as it has been completely masked (eliminated).

Following this interpretation guideline, we translate the network into the following rules:

$$\text{IF } \left[x_{2|0.7}\right]_{0.9} \text{ or } \left[x_{2|0.1} \text{ and } x_{1|0.3}\right]_{0.4} \text{ THEN } y.$$

Note that the numbers associated with the individual rules and their conditions serve as annotations (quantifications) of the corresponding components.

While the above guidelines are useful in a systematic enumeration of the rules residing within the network (and owing to the logical underpinnings of its architecture), the interpretation of larger networks still may require more attention and calls for a substantially higher effort. It would be beneficial to eliminate the most irrelevant (insignificant) segments of the network before proceeding with the enumeration of the rules. Let us introduce two schemes supporting a methodical reduction of the network. In the first one, we retain the most significant connections. In the second method, we convert the connections of the networks into their Boolean (two-valued) counterparts.

In light of the comments about the relevance of the connections, depending on the type of the neuron, we reduce the weakest connections to 0 or 1. This is done by introducing some threshold mappings with the values of thresholds (λ and μ, respectively) coming from the unit interval. For the OR neuron, we consider the following reduction of the connections ϕ_λ: $[0, 1] \rightarrow [\lambda, 1] \cup \{0\}$ such that

$$\varphi_\lambda(w) = \begin{cases} w & \text{if } w \geq \lambda \\ 0 & \text{if } w < \lambda \end{cases}$$

Hence any connection whose value is lower than some predefined threshold λ becomes reduced to zero (therefore the corresponding input gets fully eliminated) while the remaining connections are left intact.

For the AND neuron, the transformation $\psi_\mu : [0, 1] \rightarrow [0, \mu] \cup \{1\}$ reads as

$$\psi_\mu(w) = \begin{cases} 1 & \text{if } w > \mu \\ w & \text{if } w \leq \mu \end{cases}$$

The connections with values greater than m are eliminated by making them equal to 1.

The connections are converted into the Boolean values. We use some threshold values λ and μ. For the OR neuron, if the connections whose values are not lower than the threshold are elevated to 1; the remaining ones are reduced to 0. For the AND neuron, the binarization of the connections is realized as follows: if the connection is higher than the threshold, it is made equal to 1, otherwise we make it equal to 0 meaning that the corresponding input is fully relevant. More formally, the transformations read as follows:

OR neuron

$$\phi_\lambda : \quad [0, 1] \rightarrow \{0, 1\}$$
$$\varphi_\lambda(w) = \begin{cases} 1 & \text{if } w \geq \lambda \\ 0 & \text{if } w < \lambda \end{cases}$$

AND neuron

$$\psi_\mu : [0, 1] \rightarrow \{0, 1\}$$
$$\psi_\mu(w) = \begin{cases} 1 & \text{if } w > \mu \\ 0 & \text{if } w \leq \mu \end{cases}$$

The choice of the threshold values implies a certain number of connections being reduced. Higher values of λ and lower values of μ give rise to more compact form of the resulting network. While being more readable and interpretable, the modified networks come with the lower approximation capabilities. Consider the data D (denoting a training or testing data set) being originally used in the development of the network. The performance index of the reduced network is typically worse than the originally developed network. The increase in the values of the performance index can be sought as a sound indicator guiding a process of forming a sound balance between the improvement in the transparency (achieved reduction) and accuracy deterioration.

For the referential neurons $y = \text{REF}(\mathbf{x};\mathbf{w},\mathbf{a})$ the pruning mechanisms may be applied to the AND neuron combining the partial results of referential computing as well as the points of reference. Considering that we are concerned with the AND neurons performing the aggregation, the connections higher than the assumed threshold are practically eliminated from the computing. Apparently we have $(w_i \, sx_i)tA \approx 1 \quad t = A$ where A denotes the result of computing realized by the neuron for the rest of its inputs. The reference point a_i requires different treatment depending upon the type of the specific referential operation. For the inclusion operation, $INCL(x, a_i)$ we can admit the threshold operation to come in the

$$INCL^\sim(x, a_i) = \begin{cases} INCL(x_i a_i) & \text{if } a_i \leq \mu \\ 1 - x & \text{if } a_i > \mu \end{cases}$$

with μ being some fixed threshold value. In other words, we consider that $INCL(x, a_i)$ is approximated by the complement of x (where this approximation is implied by the interpretational feasibility rather than being dictated by any formal optimization problem), $INCL(x, a_i) \approx 1 - x$. For the dominance neuron we have the expression for the respective binary version of $DOM, DOM^\sim$:

$$DOM^\sim(x, a_i) = \begin{cases} DOM(x_i, a_i) & \text{if } a_i \leq \mu \\ x & \text{if } a_i > \mu \end{cases}$$

The connection set up to 1 is deemed essential. If we accept a single threshold level of 0.5 and apply this consistently to the all the connections of the network and set up the threshold 0.1 for the inclusion neuron, the statement

$$y = [x_1 \; included \; in \; 0.6] \; or \; 0.2 \; and \; [x_2 \; included \; in \; 0.9] \; or \; 0.7$$

Translates into a concise (yet approximate) version assuming the form of the following logic expression:

$$y = [x_1 \; included \; in \; 0.6]$$

The choice of the threshold value could be a subject of a separate optimization phase but we can also admit some arbitrarily values especially if we are focused on the interpretation issues.

3.4.7 Learning Algorithms in Logic-Oriented Networks

The learning of the logic-oriented networks benefits greatly from the incorporation of some prior domain knowledge which is seamlessly used to shape up the structure of the network and establish some the most meaningful connections. This unique feature is not available in case of "standard" neural networks as they do not come with any direct interpretability capabilities which could be instantaneously taken advantage of when some pieces of domain knowledge become available. Having said that, all learning mechanisms that are available within the realm of neuro-computing are equally applicable and useful in the setting of logic-oriented networks. As usual, two facets of learning are sought that is supervised and unsupervised learning schemes. Both structure and parameters of the network could be optimized. In case of structural learning, the critical aspects to be taken into account involve: (a) the number of nodes in the hidden layer, (b) types of referential operations realized by the individual logic neurons, (c) choice of a certain T-norm and T-conorm, and (d) determination of the neutral values of the uninorms in case of unineurons used in the logic network. The structural optimization is carried out by using techniques of evolutionary optimization. Quite commonly genetic algorithms, particle swarm optimization, ant colonies, evolutionary programming are of interest in this framework (Markowska-Kaczmar and Trelak 2005; Michalewicz 1996). The essential aspect here is to encode the structure so that all components determining it are represented.

The parametric optimization is a domain of gradient-based learning where the parameters (connections, reference points, etc.) are subject to adjustments. In general, the modifications are governed by the gradient of the objective function (performance index) computed with respect to the individual parameters of the network. Here comes an interesting learning strategy. Considering that some parts of the network have been established on a basis of prior knowledge, we may set the corresponding values of the connections to the values resulting from the translation of this domain knowledge. The connections which are to be learned can be set to some random values. This arrangement sets up a certain starting point in the search space of all connections which is far more promising than a complete random setup we typically encounter when learning neural networks.

With regard to the learning itself, it is instructive to draw attention to the way in which a minimized performance index becomes expressed. Essentially there are two ways of doing that. In this context the role of the output interface becomes visible. The two ways of expressing the performance index are referred to as an internal and external level of network's accuracy characterization and optimization. Their essence is visualized in Fig. 3.25. At the external (viz. numeric) level of accuracy quantification, we compute a distance between the numeric data and the numeric output of the model resulting from the transformation realized by the interface of the model. In other words, the performance index expressing the (external) accuracy of the reads in the form

$$Q = \sum_{k=1}^{N} \left\| \mathbf{y}_k - \mathbf{target}_k \right\|^2 \tag{3.5}$$

where the summation is carried out over the numeric data available in the training, validation, or testing set. The form of the specific distance function (Euclidean, Hamming, Tchebyschev or more generally, Minkowski distance) could be selected when dealing with the detailed quantification of the proposed performance index.

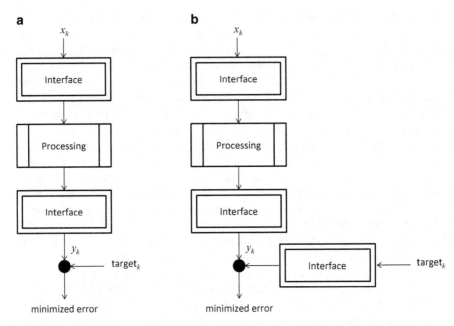

Fig. 3.25 Two fundamental ways of expressing the accuracy of fuzzy models: (**a**) at the numeric level of experimental data and results of mapping through the interface and (**b**) at the internal level of processing after the transformation through the interfaces

At the internal level of assessment of the quality of the fuzzy model, we transform the output data through the output interface so now they become vectors in the $[0.1]^m$ hypercube and calculate distance at the internal level by dealing with two vectors with the $[0, 1]$ entries. As before, the calculations may involve training, validation or testing data. More specifically, we have

$$Q = \sum_{k=1}^{N} \|\mathbf{u}_k - \mathbf{t}_k\|^2, \tag{3.6}$$

refer also to Fig. 3.25b. These two ways of quantifying the accuracy are conceptually different and there is no equivalence between them unless there granular to numeric interface does not produce any additional error. We have already elaborated on this issue in Sect. 3.3. Quite often the interface itself could introduce an additional error. In other words, we may have a zero error at the level of granular information, however, once we transform these results through the interface they become associated with the non-zero error. The performance index in the form shown above Eqs. 3.5 and 3.6 is computed either at the numeric or granular level. In the first case, refer to Fig. 3.25a, it concerns the real numbers. At the level of information granules, the distances are determined at the level of the elements located in the unit hypercubes $[0, 1]^m$, see Fig. 3.25b.

Table 3.1 Comparison of fuzzy systems and traditional neural networks

Knowledge representation	Fuzzy systems	Neural networks
Uncertainty & imprecision tolerance	+ (good)	+ (good)
Generalization ability	+ (good)	+ (good)
Training ability/adaptability	− (bad)	+ (good)
Knowledge transparency/explanation ability	+ (good)	− (bad)
Knowledge discovery	− (bad)	+ (good)

3.4.8 Neuro-Fuzzy Systems

Both neural networks and fuzzy systems have some things in common. They can be used for solving a problem (e.g. pattern recognition, regression or density estimation) if there does not exist any mathematical model of the given problem.

Traditional (non-logic oriented) neural networks can only come into play if the problem is expressed by a sufficient amount of observed examples. These observations are used to train the black box. On the one hand no prior knowledge about the problem needs to be given. On the other hand, however, it is not straightforward to extract comprehensible rules from the neural network's structure.

Fuzzy systems are more user-friendly as they utilize human-like knowledge representation in contrast to traditional ANNs which seem opaque black-box systems to the user. On the contrary, a fuzzy system demands linguistic rules instead of learning examples as prior knowledge. Furthermore the input and output variables have to be described linguistically. If the knowledge is incomplete, wrong or contradictory, then the fuzzy system must be tuned. Since there is not any formal approach for it, the tuning is performed in a heuristic way. This is usually very time consuming and error-prone. We can say, in general, fuzzy systems lack the ability to learn from examples for a new behavior.

The rationale behind combination of both concepts inside a single neuro-fuzzy system is preserving their advantages while removing their disadvantages when being used solely (see Table 3.1).

In capacity of a neuro-fuzzy system, the fuzzy neural network becomes more logic-transparent and gains the abilities to accept, train and extract a robust set of fuzzy rules.

As a type of multi-layer neural network, a neuro-fuzzy system has input and output layers and three or more hidden layers that represent fuzzy rules and membership functions (for the terms used in the rules).

Neuro-fuzzy systems can implements different fuzzy rule based logic schemes. Most widely used type-1 fuzzy networks are simulations of the inference systems for Mamdani and Takagi-Sugeno-Kang (TSK) fuzzy If-Then rules. A version of the latter is also known under the name of Adaptive Neuro-Fuzzy Inference System (ANFIS). More complex versions of neuro-fuzzy systems can implement versions of type-2 fuzzy logic inference schemes (Hidalgo et al. 2009).

In general, the neuro-fuzzy system simulates the behavior of a system based on If-Then rules exploiting fuzzy and type-2 fuzzy linguistic terms (Huang and Wu 2009). The linguistic terms are implemented in a network as objects with adjustable properties. The whole set or adjustable properties form the set of adjustable parameters W. For example, a neuron implementing a type-2 fuzzy term as trapezoidal membership function can have six parameters (see Sect. 3.6).

3.4.9   Type-1 Fuzzy Mamdani Logical Neural Network

The Mamdani fuzzy inference model was proposed for generating fuzzy rules from input–output data (Mamdani and Assilian 1975; Mamdani 1977). A typical Mamdani rule can be expressed as follows:

$$R^i : \text{If } x_1 \text{ is } \widetilde{T}^i_1 \text{ and } x_2 \text{ is } \widetilde{T}^i_2 \text{ and } \dots x_s \text{ is } \widetilde{T}^i_s \text{ Then } y \text{ is } \widetilde{T}^i_{s+1},$$

where x_j ($j = 1, 2, \dots, s$) and y are input and output variables, respectively; $\widetilde{T}^i_j \in \mathbf{T}_j$ ($j = 1, 2, \dots, s$) is a antecedent linguistic term (fuzzy set) from the term-set (codebook) $\mathbf{T}_j$ used for variable x_j in rule i and $\widetilde{T}^i_{s+1} \in \mathbf{T}_{s+1}$ is a consequent fuzzy set from $\mathbf{T}_{s+1}$ used for output variable y in rule i.

For example, let's consider the rules (Negnevitsky 2005):

$$R^1 : \text{IF } x_1 \text{ is } \widetilde{A}_1 \text{ and } x_2 \text{ is } \widetilde{B}_1 \text{ THEN } y \text{ is } \widetilde{C}_2,$$

$$R^2 : \text{IF } x_1 \text{ is } \widetilde{A}_1 \text{ and } x_2 \text{ is } \widetilde{B}_2 \text{ THEN } y \text{ is } \widetilde{C}_2,$$

$$R^3 : \text{IF } x_1 \text{ is } \widetilde{A}_1 \text{ and } x_2 \text{ is } \widetilde{B}_3 \text{ THEN } y \text{ is } \widetilde{C}_1,$$

$$R^4 : \text{IF } x_1 \text{ is } \widetilde{A}_2 \text{ and } x_2 \text{ is } \widetilde{B}_2 \text{ THEN } y \text{ is } \widetilde{C}_2,$$

$$R^5 : \text{IF } x_1 \text{ is } \widetilde{A}_3 \text{ and } x_2 \text{ is } \widetilde{B}_3 \text{ THEN } y \text{ is } \widetilde{C}_2,$$

$$R^6 : \text{IF } x_1 \text{ is } \widetilde{A}_3 \text{ and } x_2 \text{ is } \widetilde{B}_1 \text{ THEN } y \text{ is } \widetilde{C}_1,$$

Figure 3.26 illustrates a possible architecture for neuro-fuzzy system simulating the inference system based on the above rules (Negnevitsky 2005).

Each layer in the neuro-fuzzy system simulates a particular operation in the fuzzy logical inference process.

The first (input) layer accepts and distributes numerical (crisp) input signals to the next layer.

The second layer performs the fuzzification procedure. The associated neurons store characteristics of the linguistic terms (fuzzy sets) used in antecedents of fuzzy rules. Every neuron in this layer determines the degree to which the input signal belongs to the associated fuzzy set.

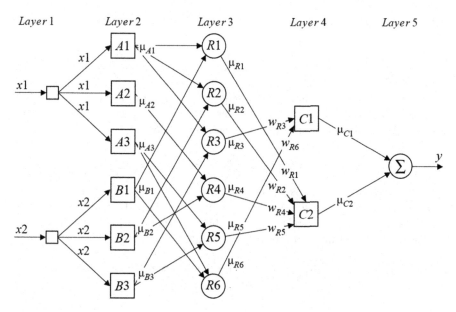

Fig. 3.26 Example architecture of a neuro-fuzzy system

The activation function of the neurons at this layer (called also fuzzifiers) is set to a function that specifies the corresponding fuzzy set, i.e. the membership function. Most frequently for this purpose an S-shaped function is used (e.g. the Gaussian function). The triangular and trapezoidal fuzzy number models are also used if the differentiability of the function is not required (like in case of evolutionary algorithm based training methods). The fuzzifiers determines the required linguistic term after appropriately setting their parameters. For a symmetric triangular fuzzy number model to approximate linguistic terms in the rules, two parameters (a and b) per a neuron are sufficient:

$$
y_i^{(1)} = \begin{cases} 0, & \text{if } x_i^{(1)} \leq a - \dfrac{b}{2} \\[2ex] 1 - \dfrac{2\left|x_i^{(1)} - a\right|}{b}, & \text{if } a - \dfrac{b}{2} < x_i^{(1)} < a + \dfrac{b}{2} \\[2ex] 0, & \text{if } x_i^{(1)} \geq a + \dfrac{b}{2} \end{cases}
$$

The neurons of third layer simulates the fuzzy rules. A fuzzy rule neuron receives inputs from the fuzzifier neurons representing linguistic terms in antecedents of the rule.

A frequently used T-norm operator used in neuro-fuzzy systems is the product operator. For the network depicted in Fig. 3.26, the output of a rule neuron R1 which corresponds to rule 1 is computed as product of inputs from neurons A1 and B2:

$$\mu_{R1} = \mu_{A1} \times \mu_{B1}$$

The next, fourth layer simulates consequents of fuzzy rules and is composed of neurons representing output linguistic terms. The T-conorm operator can be implemented as the probabilistic OR operator (Fig. 3.26):

$$\mu_{C1} = \mu_{R3} \oplus \mu_{R6}$$

The outputs of the neurons on fourth layer represent output fuzzy sets.

Finally, at fifth layer, the defuzzification process is performed. Neuro-fuzzy systems can use any of a number of defuzzification technique including the Center of Gravity (COG) or the sum-product composition:

$$y = \frac{\mu_{C1} \times a_{C1} \times b_{C1} + \mu_{C2} \times a_{C2} \times b_{C2}}{\mu_{C1} \times b_{C1} + \mu_{C2} \times b_{C2}}$$

3.4.10   Type-1 Fuzzy TSK Logical Neural Network

The TSK fuzzy inference model was proposed for generating fuzzy rules from input–output data (Takagi and Sugeno 1983). A typical TSK rule can be expressed as follows:

$$R^i : \text{ IF } x_1 \text{ is } \widetilde{T}_1^i \text{ and } x_2 \text{ is } \widetilde{T}_2^i \text{ and } \ldots x_s \text{ is } \widetilde{T}_s^i \text{ THEN } y \text{ is } f(x_1, x_2, \ldots, x_s)$$

where $x_1, x_2, \ldots, x_n$ are input variables, $\widetilde{T}_j^i$ are fuzzy sets, f is crisp a (usually polynomial) function of the inputs. When y is a constant the model is called a zero order TSK model, when it is a first order polynomial, it is called a first order TSK model:

$$y = k_0 + k_1 x_1 + \ldots + k_s x_s$$

Figure 3.27 illustrates an example architecture for TSK logical neural network (Negnevitsky 2005).

Neurons of the first layer simply distribute crisp signals to the next, fuzzification, layer. Usual type of membership function is the bell-shaped activation function. The third layer represents rules. Firing strengths are calculated, which are outputs from neurons at this layer. The product operator is frequently used as conjunction (T-norm) for this purpose. For the network shown in Fig. 3.27, the firing strength would be:

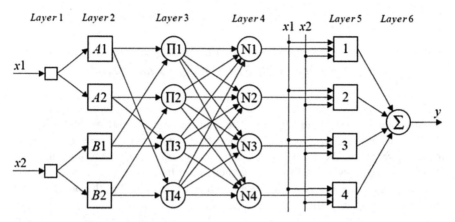

Fig. 3.27 Example of TSK logical neural network

$$y_{\Pi_1} = \mu_{A1} \times \mu_{B1}$$

Next, fourth layer is the normalization layer, where the neurons calculates the normalized firing strength for a particular rule. It represents the contribution of a rule to the final output and is computed as the ratio of the firing strength of the considered rule to the sum of firing strengths of all rules:

$$y_{N1} = \frac{\mu_1}{\mu_1 + \mu_2 + \mu_3 + \mu_4}$$

The fifth layer's neurons perform the defuzzification:

$$y_i^{(5)} = x_i^{(5)}(k_{i0} + k_i x_1 + k_{i2} x_2)$$

Note that the neurons at this layer receive inputs not only from the previous layer but also from original crisp inputs (in this case x_1 and x_2).

Final sixth layer neuron perform just summation of all of its input signals:

$$y = \sum_{i=\overline{1,4}} x_i^{(6)}$$

3.5   Type-1 Fuzzy Neural Networks and Their Learning

3.5.1   *Gradient Descent Based Training Algorithm for Fuzzy Neural Networks with Fuzzy Weights and Signals*

Let's consider training of a layered neural network with fuzzy signals and parameters (weights and thresholds) by using the gradient-descent based method. Without limiting the generality, for simplicity we consider symmetric triangular fuzzy model to represent the fuzzy signals and parameters. Also we consider that the network is feed-forward with sigmoidal activation function for neurons at hidden and output layers, has one hidden layer, and a single neuron at the output layer, i.e. the architecture is $A[n_{input}, n_{hidden}, 1]$.

Consider that the training set consists of P input–output vectors with fuzzy values:

$$\left(\mathbf{X}_p^*, y_p^* \right), \quad p = 1, \ldots P$$

The outputs of each neuron are computed according to the formula

$$\widetilde{y}_i^{(1)} = f\left(\left(\sum_{j=1, n_{input}} \widetilde{w}_{i,j}^{(1)} \widetilde{x}_j \right) + \widetilde{\theta}_i^{(1)} \right), \quad i = 1, \ldots, n_{hidden}$$

$$\widetilde{y}^{(2)} = f\left(\left(\sum_{j=1, n_{hidden}} \widetilde{w}_j^{(2)} \widetilde{y}_j^{(1)} \right) + \widetilde{\theta}^{(2)} \right)$$

Therefore, the sets of adjustable parameters are:

$$\mathbf{w}^{(1)} \equiv \begin{pmatrix} \widetilde{w}_{1,1}^{(1)} & \widetilde{w}_{1,2}^{(1)} & \cdots & \widetilde{w}_{1,n_{input}}^{(1)} & \widetilde{\theta}_1^{(1)} \\ \widetilde{w}_{2,1}^{(1)} & \widetilde{w}_{2,2}^{(1)} & \cdots & \widetilde{w}_{2,n_{input}}^{(1)} & \widetilde{\theta}_2^{(1)} \\ \cdots & \cdots & \cdots & \cdots & \cdots \\ \widetilde{w}_{n_{hidden},1}^{(1)} & \widetilde{w}_{n_{hidden},2}^{(1)} & \cdots & \widetilde{w}_{n_{hidden},n_{input}}^{(1)} & \widetilde{\theta}_{n_{hidden}}^{(1)} \end{pmatrix}$$

$$\mathbf{w}^{(2)} \equiv \begin{pmatrix} \widetilde{w}_{1,1}^{(2)} & \widetilde{w}_{1,2}^{(2)} & \cdots & \widetilde{w}_{1,n_{hidden}}^{(2)} & \widetilde{\theta}_1^{(2)} \end{pmatrix}$$

Since all parameters are symmetric triangular numbers, each of them can be represented by two real numbers as follows:

$$\widetilde{w}_{i,j} = T\left(\left[w_{i,j} \right]^L, \left[w_{i,j} \right]^R \right) = T\left(\left[w_{i,j} \right]^L, \frac{\left[w_{i,j} \right]^L + \left[w_{i,j} \right]^R}{2}, \left[w_{i,j} \right]^R \right)$$

Then, considering α-level sets, the outputs, targets, and weights are represented as follows:

$$\left[y_p\right]^{(\alpha)} = \left[\widetilde{y}^{(2)}\left(\mathbf{X}_p^*\right)\right]^{(\alpha)} = \left[\left[y_p\right]^L(\alpha), \left[y_p\right]^R(\alpha)\right]$$

$$\left[y_p^*\right]^{(\alpha)} = \left[\left[y_p^*\right]^L(\alpha), \left[y_p^*\right]^R(\alpha)\right]$$

$$\left[w_{i,j}\right]^{(\alpha)} = \left[\left[w_{i,j}\right]^L(\alpha), \left[w_{i,j}\right]^R(\alpha)\right]$$

Since f is strictly monotone increasing we have:

$$\left[\widetilde{y}_i^{(1)}\right]^{(\alpha)}$$
$$= \left[f\left(\left(\sum_{j=1,n_{input}} \left[\widetilde{w}_{i,j}^{(1)}\widetilde{x}_j\right]^L\right) + \left[\widetilde{\theta}_i^{(1)}\right]^L\right), f\left(\left(\sum_{j=1,n_{input}} \left[\widetilde{w}_{i,j}^{(1)}\widetilde{x}_j\right]^R\right) + \left[\widetilde{\theta}_i^{(1)}\right]^R\right)\right],$$
$$i = 1, \ldots, n_{hidden}$$

$$\left[\widetilde{y}^{(2)}\right]^{(\alpha)} =$$
$$\left[f\left(\left(\sum_{j=1,n_{hidden}} \left[\widetilde{w}_j^{(2)}\widetilde{y}_j^{(1)}\right]^L\right) + \left[\widetilde{\theta}^{(2)}\right]^L\right), f\left(\left(\sum_{j=1,n_{hidden}} \left[\widetilde{w}_j^{(2)}\widetilde{y}_j^{(1)}\right]^R\right) + \left[\widetilde{\theta}^{(2)}\right]^R\right)\right]$$

The error function can be expressed as follows:

$$E = \sum_p E_p$$

$$E_p = \sum_\alpha \alpha \cdot e_p(\alpha)$$

$$e_p(\alpha) \equiv \frac{1}{2}\left(\left(\left[y_p^*\right]^L(\alpha) - \left[y_p\right]^L(\alpha)\right)^2 + \left(\left[y_p^*\right]^R(\alpha) - \left[y_p\right]^R(\alpha)\right)^2\right)$$

Then:

$$\left[\Delta w_{ij}^{(l)}\right]^L = -\eta \frac{\partial e_p(\alpha)}{\partial \left[w_{ij}^{(l)}\right]^L}$$

$$\left[\Delta w_{ij}^{(l)}\right]^R = -\eta \frac{\partial e_p(\alpha)}{\partial \left[w_{ij}^{(l)}\right]^R}$$

The explicit calculation of above derivatives can be found in (Ishibuchi et al. 1995, pp. 291–292). Finally, we have the fuzzy weight update as (Fuller 2000):

$$\widetilde{w}_{i,j}(t+1) \equiv T\left(\left[w_{i,j}(t+1)\right]^{L}, \left[w_{i,j}(t+1)\right]^{R} \right)$$

$$\left[w_{i,j}(t+1)\right]^{L} = \min\left(\left[\widetilde{w}_{ij}^{(l)}\right]^{L} + \left[\Delta w_{ij}^{(l)}\right]^{L}, \left[\widetilde{w}_{ij}^{(l)}\right]^{R} + \left[\Delta w_{ij}^{(l)}\right]^{R} \right)$$

$$\left[w_{i,j}(t+1)\right]^{R} = \max\left(\left[\widetilde{w}_{ij}^{(l)}\right]^{L} + \left[\Delta w_{ij}^{(l)}\right]^{L}, \left[\widetilde{w}_{ij}^{(l)}\right]^{R} + \left[\Delta w_{ij}^{(l)}\right]^{R} \right)$$

As with the crisp case, before starting training, all weights should be initialized by random fuzzy numbers. The above adjustments should be cyclically done for all considered α-levels and for all training data p. Training stops when the error level E is below an acceptable maximum level $E_{\max}$.

3.5.2   DE Based Training Algorithm for Fuzzy Feed-Forward and Recurrent Neural Networks

As we have already mentioned, training of pure fuzzy neural networks (FNN) is very complex and time-consuming task. In most cases the training requires a global optimization method suitable for nonlinear, non-convex, and non-differentiable functions (problems).

The training of FNN is further complicated when we deal with applications of temporal character such as dynamic control, forecasting, identification, recognition of temporal sequences (e.g. voice recognition). It is obvious that in this case classical FNN with feed-forward structure lacking the effect of memorizing of signal series could be ineffective. In this respect there is a strong demand for recurrent fuzzy neural networks (RFNN) capable of retrieving dynamic models defining fuzzy input/output relationship by means of training on the basis of example data represented as fuzzy input/output time series or fuzzy if-then rules. This ability makes them very useful in solving temporal problems.

Despite the fact that the gradient descent based methods are predominant, they are not global optimizers. Most suitable are population based optimization techniques including genetic algorithms, evolutionary strategy, particle swarm optimization, DE, etc. We suggest to use DE method (Price et al. 2005) which has many advantages over other evolutionary algorithms and GA (Aliev et al. 2009).

There are some reasons for using DE in FNN training problem. First, DE supports a search mechanism of global nature. DE is useful when dealing with different distance functions including Hamming distance, Tschebyshev distance (gradient descent based methods require distance functions be differentiable, e.g. Euclidean distance function).

The problem of learning of fuzzy neural networks is an optimization problem of adjusting fuzzy parameter vectors $\widetilde{W} = \left\{\widetilde{w}_{lij}\right\}$ and $\widetilde{V} = \left\{\widetilde{v}_{lij}\right\}$ to minimize the error function

$$E = \frac{1}{P \cdot D} \sum_p \sum_i d\left(\widetilde{y}_{pi}, \widetilde{y}_{Npi}\right)$$

where $\widetilde{y}_{pi}$ is the desired value and $\widetilde{y}_{Npi}$ is the actual value of RFNN output layer's neuron i when applied p-th training vector, P is the number of training input/output vectors and D is the dimension of output vectors, $d\left(\widetilde{y}_{pi}, \widetilde{y}_{Npi}\right)$ is the distance between vectors $\widetilde{y}_{pi}$ and $\widetilde{y}_{Npi}$.

As a measure of distance between $\widetilde{y}(t)$ and $\widetilde{y}_N(t)$, one from the long list of distance functions (Aliev and Aliev 2001; Pedrycz and Reformat 2005) can be used. For example, the fuzzy distance between fuzzy numbers $\widetilde{A} = \left\{\left[A_L^{(\alpha)}, A_R^{(\alpha)}\right]/\alpha\right\}$ and $\widetilde{B} = \left\{\left[B_L^{(\alpha)}, B_R^{(\alpha)}\right]/\alpha\right\}$ can be computed as follows

$$\widetilde{d}(A,B) = \left\{\left[\left|A_L^{(\alpha)} - B_L^{(\alpha)}\right|, \left|A_R^{(\alpha)} - B_R^{(\alpha)}\right|\right]/\alpha\right\}$$

The defuzzified value of the distance would be:

$$d(A,B) = \int_0^1 \left(\left|A_L^{(\alpha)} - B_L^{(\alpha)}\right| + \left|A_R^{(\alpha)} - B_R^{(\alpha)}\right|\right) \alpha \, d\alpha$$

Training algorithm is critical to FNN as it will affect FNN approximation capability. Due to the reasons pointed out in the beginning of this section, to train FNN we use one of recently developed excellent performance population-based global optimizer, the differential evolution (DE) algorithm (Price et al. 2005).

During the training, the weights of feed-forward and feed-back connections and biases of FNN are changed by the DE algorithm to drive the network to a state with the error level possibly close to zero.

To apply an evolutionary population based optimization (DEO) algorithm to train a fuzzy neural network we first should identify the optimized parameter vector. Let's consider that in the networks shown in Figs. 3.10 and 3.12 all signals (i.e. neuron input and output activations) are fuzzy numbers of a chosen format, e.g. triangular symmetric fuzzy numbers $T(b, \varepsilon) \equiv T(b - \varepsilon, b, b, b + \varepsilon)$. This assumption means that two parameters will be associated with every connection weight $w_{i,j}^{(l)}$ and bias $\theta_i^{(l)}$ (threshold) value in the considered feed-forward fuzzy NN. Then, the number of all parameters to train will be twice as much as for crisp version of this network.

For a feed-forward fuzzy neural network (Fig. 3.10) this will be:

$$N_{par} = 2((n_0 + 1)n_1 + (n_1 + 1)n_2 + \ldots + (n_{L-2} + 1)n_{L-1}). \qquad (3.7)$$

And for a recurrent fuzzy network (Fig. 3.12) even more:

$$N_{par} = 2\Big(\big[[(n_0+1)n_1 + (n_1+1)n_2 + \ldots + (n_{L-2}+1)n_{L-1}\big] + \big[n_1^2 + n_2^2 + \ldots + n_{L-1}^2\big]\Big) \tag{3.8}$$

Without limiting generality let's consider that our fuzzy neural network consists of three layers (which is quite sufficient for most of problems). Let's assume that the architecture is $A[N_i, N_h, N_o]$, which means that the input, hidden, and output layers have numbers of neurons N_i, N_h, N_o, respectively.

Before starting training all the parameters are initialized by randomly chosen values, usually not beyond the interval $[-1, 1]$. This constraint is further enforced to the parameters associated to backward connections for recurrent networks (RNN). It means that during further training steps the values of forward weights and biases can go beyond the interval $[-1, 1]$ while the values of backward connection are kept within this interval. This additional constraint is added to make RNN stable which means that under the constant input the value of output will converge to a constant value (either crisp or fuzzy).

Prior launching the optimization we set parameter f of DE to a positive value (typically about 0.9), define the DE cost function to be the chosen error function, and choose the population size (typically ten times the number of optimization parameters, i.e. $10N_{par}$). Then the differential evolution optimization is started.

DE based RFNN training algorithm can be summarized as presented in Fig. 3.28.

If the obtained total error performance index or the behavior of the obtained network is not desired, we can restructure the network by adding new hidden neurons, or do better granulation of the learning patterns.

During the DE optimization process the solutions resulting in lower cost values have more chances to survive and be saved into a new population for participation in future generations. The process is repeated iteratively. During succeeding generations we keep into the population the solution that produced the lowest value of cost function of all previous generations. The farther we go with generations the higher is the chance to find a better solution.

3.6 Combining Type-2 Fuzzy Sets with Neural Networks

3.6.1 Introductory Notes

Neural systems are widely used today as controllers in highly dynamic and uncertain environments. The systems constructed on the base of type-1 membership functions may not always be able to efficiently handle the cognitive uncertainties (i.e. uncertainties about the correctness of the results of the own information processing) associated with complex processes (Aliev et al. 2011; Abiyev et al. 2013).

Step 0. Initialize DE
        Step 0.0.   Define the structure of FNN: N_i, N_h, N_o
        Step 0.1.   Construct template parameter vector X of dimension N_{par} according
                               (3.8) for holding RFNN weights and biases:
$$X = \{ \tilde{W}, \tilde{V}, \tilde{\theta} \}$$
        Step 0.2.   Set algorithm parameters:
                               f (mutation rate),
                               cr (crossover rate), and
                               $PopSize$ (size of population)
        Step 0.3.   Define the cost function as function of error function of current
                               RFNN parameters: $\tilde{E} = \sum \sum d\left(\tilde{y}_{pi}, \tilde{y}_{Npi} \right)$

Step 1.   Randomly generate $PopSize$ parameter vectors (from respective parameter
                       spaces (e.g. in the range [-1, 1]) and form a population $P = \{X_1, X_2, ..., X_{ps}\}$
Step 2.   While Termination condition (number of predefined generations reached or
                       required error level obtained) is not met generate new parameter sets:
        Step 2.1.   Choose a next vector X_i ($i=1,..., PopSize$)
        Step 2.2.   Choose randomly different 3 vectors from P:
                               X_{r1}, X_{r2}, X_{r3} each of which is different from current X_i
        Step 2.3.   Generate trial vector $X_t = X_{r1} + f(X_{r2} - X_{r3})$
        Step 2.4.   Generate new vector from trial vector X_t. Individual vector
                               parameters of X_t are inherited with probability cr into the new vector
                               X_{new}. If the cost function from X_{new} is better (lower) than the cost
                               function from X_i, current X_i is replaced in population P by X_{new}
      Next i
Step 3.   Select the parameter vector X_{best} (NN parameter set) with best cost (training
                     error $\tilde{E}$) function from population P. Extract from X_{best} parameters vectors
                     ($\tilde{W}, \tilde{V}$) defining weights and thresholds for FNN
Step 4.   Save the parameters.

Fig. 3.28 DE based RFNN training algorithm

To enable a system to deal with cognitive uncertainties in a manner more like humans do, the concept of type-2 fuzzy logic can be incorporated into neural networks (Karnik and Mendel 1999).

A type-2 fuzzy logical neural system is constructed similarly to type-1 system by integrating type-2 fuzzy sets, the inference of type-2 fuzzy logic system, and a neural network architecture. The system uses IF–THEN rules to provide a mapping from input type-2 fuzzy sets to output type-2 fuzzy sets (Karnik and Mendel 1999).

As a more advanced version of neuro-fuzzy system, it may effectively adopt the knowledge from experts, when available, to form an initial rules base, but can also generate it from raw input-to-output data patters. A greater advantage, however, over traditional (not network based) inference engines is that the knowledge base (the provided set of rules) is adaptable based on low level experimental input-to-output data.

Considering higher computational burden dealing with pure type-2 fuzzy sets, in most cases, when implementing type-2 fuzzy logical neural networks, interval fuzzy membership functions are used to describe variable values used in IF-THEN rules. Note that an interval type-2 fuzzy logical neural system is proven to be a universal function approximator (Castillo et al. 2013; Castro et al. 2009).

The following sections consider examples of neural network based type-2 fuzzy logical inference systems. The Chap. 4 will consider clustering technique, which can be used to generate IF-THEN rules for type-2 fuzzy neural systems without involving the experts.

3.6.2 Type-2 Fuzzy Logical Neural Networks (T2FLNN)

A type-2 fuzzy logical neural network (T2FLNN) implements type-2 fuzzy logic system and some of their parameters and components are presented by fuzzy logic terms (Aliev et al. 2011; Wang et al. 2004). Successive layers of the network perform type-2 fuzzification, represent fuzzy rules, define the consequences of each rule, and realize an aggregation of the type-2 fuzzy (or type-1 fuzzy) output values for each output variable. Finally, the defuzzification takes place. T2FLNN can effectively represent fuzzy logic schemes (FLS) with type-2 fuzzy If-Then rules.

The architecture of an example T2FLNN (Aliev et al. 2011) is shown in Fig. 3.29. Let us elaborate on the functionality of the system's six layers in more detail. Layer 0 (*Inputs*) is the input layer accepting crisp input signals. Layer 1 (*Input MFs*) consists of fuzzifiers that map inputs to type-2 fuzzy terms used in the rules. Layer 2 (*Rules*) comprises nodes representing these rules. Each rule node performs the Min operation on the outputs (interval valued membership degrees) of the incoming links from the previous layer. Layer 3 (*Output MFs*) consists of output terms membership functions of type-1. Layer 4 (*Aggregation*) computes the fuzzy output signal for the output variables. Layer 5 (*Defuzzification*) realizes the defuzzification using the Center-of-Gravity (COG) defuzzification.

3.6.3 Fuzzification and Inference Procedure of T2FLNN

The considered T2FLNN (Aliev et al. 2011) uses an arbitrary number of type-2 fuzzy input variables and an arbitrary number of type-1 fuzzy output variables. The input variable's type-2 fuzzy terms are described as:

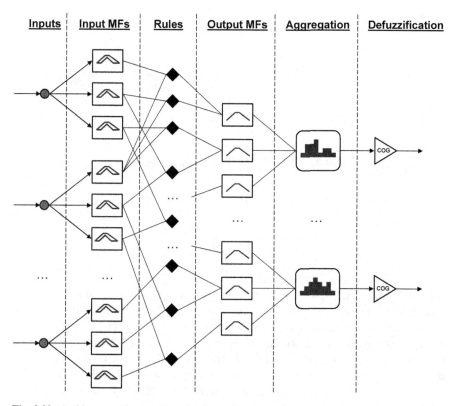

Fig. 3.29 Architecture of example type-2 fuzzy neural network (T2FLNN)

$$\widetilde{A} = \{x/\mu_x\}, \quad x \in U \subset \mathfrak{R},$$
$$\mu_x = \{\alpha/1\}, \quad \alpha \in M_x \subset [0,1],$$

where:

$$M_x = \begin{cases} [\alpha_{x1}, \alpha_{x2}], & \text{if} \left(\dfrac{\alpha_{x1} + \alpha_{x2}}{2} < \dfrac{\alpha_{x3} + \alpha_{x4}}{2}\right) \\ [\alpha_{x3} + \alpha_{x4}], & \text{elsewise} \end{cases}$$

and

$$\alpha_{x1} = \max\left(\min\left(1, \frac{RL - x}{RL - ML}\right), 0\right),$$

Fig. 3.30 A type-2 fuzzy
term value

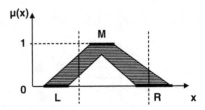

$$\alpha_{x2} = \max\left(\min\left(1, \frac{RR - x}{RR - MR}\right),\ 0\right),$$

$$\alpha_{x3} = \max\left(\min\left(1, \frac{x - LL}{ML - LL}\right),\ 0\right),$$

$$\alpha_{x4} = \max\left(\min\left(1, \frac{x - LR}{MR - LR}\right),\ 0\right).$$

In the above equations LL, LR, ML, MR, RL, RR ($LL \leq LR \leq ML \leq MR \leq RL \leq RR$) are parameters defining the "shape" of the type-2 fuzzy membership functions. An example of type-2 fuzzy value defined using this type-2 fuzzy number representation ([0.25, 0.75], [1.25, 1.75], [2.25, 3.00]) is shown in Fig. 3.30. As it can be seen, this type-2 fuzzy number is composed of three intervals, $[LL,LR]$ (a left interval, identified by letter L in the figure), $[ML,MR]$ (a medium interval, indicated by letter M), and $[RL,RR]$ (a right interval, denoted by letter R): $[LL, LR]$, $[ML, MR]$, $[RL, RR]$. The input term membership functions can be considered as interval-valued membership functions (interval membership values for two values of x are shown: $x = 1$ and $x = 2.5$).

The output variable's type-1 fuzzy terms are type-1 fuzzy trapezoidal numbers taking on the form:

$$B = [L, ML, MR, R] = [[L, L], [ML, MR], [R, R]]$$

Zadeh's implication is used to compute the output membership functions. As a result, for every output variable, we obtain two piecewise linear membership functions:

$$\tilde{y}_i = \{y/[\mu_{Li}(y), \mu_{Ri}(y)]\}$$

Type reduction is performed on the basis of the center of gravity (COG) defuzzification procedure, that is

$$COG(\tilde{y}_i) = \{y/[COG(\mu_{Li}(y)), COG(\mu_{Ri}(y))]\} = \{y/[y_{Li}, y_{Ri}]\}$$

The final defuzzification is done through averaging the two values:

$$y_i = \frac{y_{Li} + y_{Ri}}{2}$$

In what follows, we discuss the implementation of the COG defuzzification used for type-reducing of aggregated output type-2 fuzzy values (with piecewise linear membership functions).

3.6.4 COG Defuzzification for Fuzzy Sets with Piecewise Linear Membership Functions

Fuzzy sets with piecewise membership functions can be produced, for example, as a result of fuzzy inference when fuzzy terms with trapezoidal membership functions are used in both in the antecedent and consequent parts of fuzzy "If-Then" rules

$$A = \{x_k/\alpha_k\}, \ k = 1, \ 2, \ \dots, \ K,$$
$$\alpha_k = \mu_A(x_k).$$

By definition:

$$A_{defuzz} = \frac{\int\limits_{x_1}^{x_K} x\mu_A(x)dx}{\int\limits_{x_1}^{x_K} \mu_A(x)dx}.$$

Any (linear) part of $\mu_A(x)$ in an interval $[x_k, x_{k+1}]$ can be represented as:

$$\mu_A(x) = \frac{\alpha_{k+1} - \alpha_k}{x_{k+1} - x_k}x + \left(\alpha_k - \frac{\alpha_{k+1} - \alpha_k}{x_{k+1} - x_k}x_k\right).$$

Then, the above integrals are calculated as:

$$\int\limits_{x_k}^{x_{k+1}} \mu_A(x)dx = \frac{(\alpha_k + \alpha_{k+1})(x_{k+1} - x_k)}{2},$$

$$\int\limits_{x_k}^{x_{k+1}} x\mu_A(x)dx = \frac{(2\alpha_k x_k + 2\alpha_{k+1}x_{k+1} + \alpha_k x_{k+1} + \alpha_{k+1}x_k)(x_{k+1} - x_k)}{6}.$$

For a single interval $[x_k, x_{k+1}]$: $A_{defuzz} = \frac{(2\alpha_k x_k + 2\alpha_{k+1}x_{k+1} + \alpha_k x_{k+1} + \alpha_{k+1}x_k)}{3(\alpha_{k+1} + \alpha_k)}.$

For the entire domain $[x_1, x_K]$ we obtain:

$$A_{defuzz} = \frac{1}{3} \frac{\sum\limits_{k=1}^{K-1} (2\alpha_k x_k + 2\alpha_{k+1}x_{k+1} + \alpha_k x_{k+1} + \alpha_{k+1}x_k)(x_{k+1} - x_k)}{\sum\limits_{k=1}^{K-1} (\alpha_k + \alpha_{k+1})(x_{k+1} - x_k)}.$$

3.7 Interval Type-2 Fuzzy Neural Network Based on TSK Logic Model

Figure 3.31 demonstrates the architecture of an interval type-2 fuzzy neural network (IT2FNN) based on the first order TSK logic model (Castillo et al. 2013). To do inference the presented IT2FNN uses a set of rules involving interval type-2 membership functions (IT2MFs) to represent linguistic terms. In the architecture presented in Fig. 3.31 rectangles are used to represent adaptive nodes and circles to represent nonadaptive nodes (Castillo et al. 2013).

The presented IT2FNN (Castillo et al 2013) consisting of seven layers is a universal approximator. The IT2FNN layers are described as follows.

Layer 0 accept real valued inputs:

$$o_i^{(0)} = x_i, i = 1, \ldots, n.$$

The next layer 1 consists of adaptive type-1 fuzzy neurons:

$$o_k^{(1)} = \mu\left(net_k^{(1)}\right),$$

where the activation function $\mu(.)$ is a membership function and

$$net_k^{(1)} = w_{k,j}^{(1)} x_i + \theta_k^{(1)}, \quad i = 1, \ldots, n; \quad k = 1, \ldots, \vartheta,$$

where $w_{k,j}^{(1)}$ and $\theta_k^{(1)}$ are adjustable parameters of each neuron – weights and thresholds, respectively.

Layer 2 contains T-norm and S-norm fuzzy neurons.

Next layer 3 computes lower and upper rule firing strengths $\left[\underline{w}^k, \overline{w}^k\right]$:

$$o_{2k-1}^{(3)} = \underline{w}^k,$$
$$o_{2k}^{(3)} = \overline{w}^k,$$
$$\underline{w}^k = \prod_{i=1}^{n} \underline{\mu}_{\tilde{F}_i^k}(x),$$
$$\overline{w}^k = \prod_{i=1}^{n} \overline{\mu}_{\tilde{F}_i^k}(x),$$

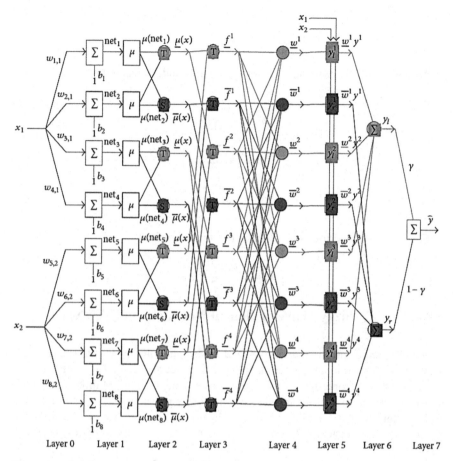

Fig. 3.31 Architecture of IT2FNN

where $\mu_{\widetilde{F}_i^k}(x) \in \left[\underline{\mu}_{\widetilde{F}_i^k}(x), \overline{\mu}_{\widetilde{F}_i^k}(x)\right]$ is the Gaussian interval type-2 membership function:

$$\mu_{\widetilde{F}_i^k}(x_i) = \exp\left(-\frac{1}{2}\left(\frac{x_i - m_i^k}{\sigma_i^k}\right)\right).$$

Layer 4 does normalization of the rule firing strength obtained in layer 3. Each output $\left[\underline{\phi}^k, \overline{\phi}^k\right]$ is computed as the ratio of k-th firing strength and the sum of the firing strengths of all rules.

Layer 5 contains neurons with adaptive parameters c_i^k. The outputs y^k are computed as follows:

$$y^k = \sum_{i=1}^{n} c_i^k x_i + c_0^k, \quad k = 1, \ldots, M.$$

At layer 6 the interval valued outputs $\tilde{y}(x)$ are produced as follows:

$$o_1^{(6)} = \tilde{y}_l(x),$$
$$o_2^{(6)} = \tilde{y}_r(x),$$
$$y_l(x) = \sum_{k=1}^{M} \underline{\phi}^k y^k,$$
$$y_r(x) = \sum_{k=1}^{M} \overline{\phi}^k y^k.$$

Finally, layer 6 performs the defuzzification. By default:

$$y(x) = \frac{y_l(x) + y_r(x)}{2}.$$

3.8 Type-2 Fuzzy Wavelet Neural Network (T2FWNN)

A wavelet neural network uses localized basis functions in the hidden layer to achieve the desired input–output mapping (Abiyev et al. 2013). The integration of the localization properties of the wavelets and the learning abilities of neural networks results in the advantages of wavelet neural networks over traditional systems for complex nonlinear system modeling (Kugarajah and Zhang 1995; Zhang and Benveniste 1992; Zhang et al. 1995; Thuillard 2001; Al-Rousana and Assaleh 2011).

A fuzzy wavelet neural network (T2FWNN) combines wavelet theory with fuzzy logic and neural networks. A combination of fuzzy technology and WNN has been efficiently used for solving signal processing and control problems (Abiyev and Kaynak 2008; Ho et al. 2001; Abiyev 2009; Sharifi et al. 2012).

Wavelet transform has the ability to analyze non-stationary signals to discover their local details. As a type-1 fuzzy system may be unable to handle uncertainties in rules and, a type-2 fuzzy system is used.

The consequent parts of the fuzzy IF-THEN rules of TSK fuzzy logical model are sometimes represented by either a constant or a linear function. Disadvantages of such systems are that they do not meet the localizability property, can model only global features of the process and may be hard to train. Also these TSK-type fuzzy networks do not provide full mapping capabilities and may need a high number of rules in order to achieve the desired accuracy for modeling complex non-linear processes. Increasing the number of the rules, however, would lead to an increase in the number of hidden neurons in the network and its overall complexity. Using wavelet rather instead of constant or linear functions can improve the computational power of the neuro-fuzzy system.

The rules in the FWNN are of the following form:

$$\text{IF } x_i \text{ is } \tilde{A}_{i1} \text{ and } x_2 \text{ is } \tilde{A}_{i2} \text{ and} \ldots \text{and } x_n \text{ is } \tilde{A}_{in} \text{ THEN } y_i \text{ is } \omega_i \sum_{j=1,..,n} \left(1 - z_{ji}^2\right) e^{-z_{ji}^2/2}$$

$$i = 1, .., r$$

where x_j, $j = 1, \ldots, n$, are (crisp) input variables, y_i, $i = 1, \ldots r$, are outputs of wavelet functions (not a defuzzified network output), A_{ij} are type-2 fuzzy linguistic terms used as value for x_j in i th rule, w_i is the weight of rule i, $z_{ji} = (x_j - b_{ji})/a_{ji}$ are parameters of the wavelet function.

The sum of wavelet output signals is calculated for each rule separately. In this way the contribution of each wavelet on the consequent part of each fuzzy rule is determined. The use of wavelets with different dilation and translation values allows the system to capture different behaviors and the essential features of the nonlinear model. The proper fuzzy model that is described by the set of IF-THEN rules can be obtained by adjusting the parameters of the membership function of the antecedent parts and the parameters of the consequent parts.

Let's now describe the operation of the type-2 fuzzy wavelet network (Abiyev et al. 2013) in more detail.

In the first layer the input signals are distributed. In the second layer each node corresponds to a linguistic term. In this layer the membership degrees are computed. Since the interval type-2 sets are used in the antecedents:

$$\mu_{\tilde{A}_k^i}(x_k) = \left[\underline{\mu}_{\tilde{A}_k^i}(x_k), \overline{\mu}_{\tilde{A}_k^i}(x_k)\right] = \left[\underline{\mu}^i, \overline{\mu}^i\right]$$

The third layer realizes the inference mechanism. In this layer using the *T-norm* product operator the membership degrees of the input signals for each rule are calculated.

$$\underline{f} = \underline{\mu}_{\tilde{A}_1^i}(x_1) * \underline{\mu}_{\tilde{A}_2^i}(x_2) * \ldots * \underline{\mu}_{\tilde{A}_n^i}(x_n)$$
$$\overline{f} = \overline{\mu}_{\tilde{A}_1^i}(x_1) * \overline{\mu}_{\tilde{A}_2^i}(x_2) * \ldots * \overline{\mu}_{\tilde{A}_n^i}(x_n)$$

The fourth layer determines the outputs of the wavelet functions in the consequent part. In the fifth layer, the output signals of the third layer are multiplied by the output signals of the wavelet functions. The contribution of each wavelet to the output of the T2FWNN is determined as follows (Fig. 3.32):

$$y_j = \omega_j \sum_{i=1}^{m} |a_{ij}|^{-1/2} \left(1 - z_{ij}^2\right) e^{-z_{ij}^2/2}$$

The sixth and the seventh layers perform the type reduction and the defuzzification operations, respectively.

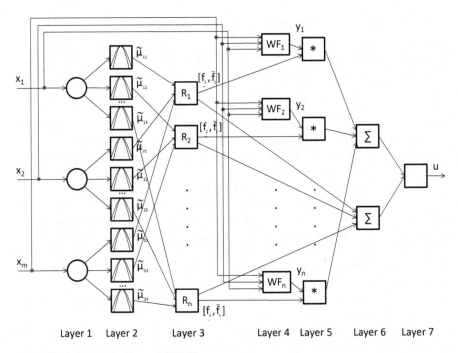

Fig. 3.32 Architecture of type-2 FWNN

The output is produced as follows:

$$u = p \frac{\sum_{j=1}^{N} f_{-j} y_j}{\sum_{j=1}^{N} f_{-j}} + q \frac{\sum_{j=1}^{N} \bar{f}_j y_j}{\sum_{j=1}^{N} \bar{f}_{-j}}$$

where N is number of active rules, p and q are design coefficients used to balance the lower and upper firing strengths of the rules.

3.9   Training of Type-2 Fuzzy Neural Networks

Training of type-2 fuzzy neural networks can be done using different methods. In case all of the neurons (including those implementing input and output fuzzifiers, type-reducers, defuzzifiers, logical operators etc.) can be represented by continuously differentiable (multi-argument and parameterized) functions and the network's architecture and inference method is not very complex, the gradient based approach can potentially be used to create a training procedure. However, for other cases, a better and even the only possible approach is an evolutionary optimization.

As we told before, the convergence of a gradient descent based strategy to adapt the parameters of the system in some cases may be faster than evolutionary. But since the former is not a global optimizer may eventually fail to train being caught at a local minimum of the error function. Failure chance may increase with large sizes of training data, complex input to output relationships, and constrained number of neurons for expressing rules and membership functions in the network, and other factors.

Below we consider in some detail an approach to training for the logical type-2 fuzzy network T2FLNN (Aliev et al. 2011) considered in Sect. 3.6.2 of this book. The gradient based methods to train IT2FNN (Sect. 3.7) and T2FWNN (Sect. 3.8) are considered in (Castillo et al. 2013) and (Abiyev et al. 2013), respectively.

Once we have obtained the initial rule base (from experts or by type-2 fuzzy clustering), we need to optimize the parameters of the rule base. To optimize these parameters a certain error function is considered as shown below

$$E = \frac{1}{n \cdot s_y} \sum_{p=1}^{n} \sum_{i=1}^{s_y} \left(y_{pi}^* - y_{pi} \right)^2.$$

Here y_{pi}^* is the desired value (target) for output i when we apply input value vector $\mathbf{x}_p$, y_{pi} is the corresponding output of the model output, n is the number of training patterns, and s_y is the number of outputs in the model (typically being equal to 1).

To arrive at the minimal value of error of the T2FLNN, the Differential Evolution (DE) algorithm (Price et al. 2005) is used. The parameters $LL, LR, ML, MR, RL,$ RR for all input terms and the parameters L, ML, MR, R for all outputs terms are adjusted.

The DE-based parameter optimization used for training of the T2FLNN (Aliev et al. 2011) can be summarized as demonstrated in Fig. 3.33.

3.10   Type-2 Fuzzy Neural Network as a Universal Approximator

Traditional neural networks, type-1 fuzzy logic systems (T1FLSs), and interval type-2 fuzzy logic systems (IT2FLSs) have been shown to be universal approximators, which means that they can approximate any nonlinear continuous function (Castro et al. 2009; Castillo et al. 2013).

It has been shown that a three-layer NN can approximate any real continuous function (Hornik et al. 1989). The same has been shown for a T1FLS (Wang and Mendel 1992a, b) using the Stone-Weierstrass theorem (Buckley 1992). Also, combining the neural and fuzzy logic paradigms (Zadeh 1965; Hirota and Pedrycz

Step 0. Initialize DE

Step 0.1.   Construct template parameter vector S of dimension (N_{par}) necessary to accommodate all parameters to be optimized. X consists of the complete set of parameters (LL, LR, ML, MR, RL, or RR) for all input terms and the complete set of parameters (L, ML, MR, or R) for all output terms.

Step 0.2.   Set the essential parameters of the DE: f (mutation rate), cr (crossover rate), and $PopSize$ (size of population).

Step 0.3.   Define the cost (objective) function expressed as the Mean Square Error (MSE) between the output of the fuzzy model with the current parameters and desired output (coming from the training data): Cost=

$$E = \frac{1}{n \cdot s_y} \sum_{p=1}^{n} \sum_{i=1}^{s_y} \left(y_{pi}^* - y_{pi}\right)^2 .$$

Step 1.   Randomly generate $PopSize$ parameter vectors (from respective parameter spaces, e.g., in the range [-1, 1]) and form a population $P = \{\mathbf{X}_1, \mathbf{X}_2, ..., \mathbf{X}_{ps}\}$.

Step 2.   While the termination condition (say, the number of predefined generations or the required error level) has not been reached, generate new parameter sets:

Step 2.1.   Choose a next vector $\mathbf{X}_i$ ($i = 1, ..., PopSize$).

Step 2.2.   Choose randomly three different vectors from P: $\mathbf{X}_{r1}, \mathbf{X}_{r2}, \mathbf{X}_{r3}$ ($1 \le r_1 \ne r_2 \ne r_3 \le PopSize$) each of which is different from the current $\mathbf{X}_i$.

Step 2.3.   Generate a trial vector $\mathbf{X}_t = (\mathbf{X}_{r1} - \mathbf{X}_{r2})f + \mathbf{X}_{r3}$.

Step 2.4.   Generate a new vector from the trial vector $\mathbf{X}_t$. Individual vector parameters of $\mathbf{X}_t$ are inherited with probability cr and included into the new vector $\mathbf{X}_{new}$. If the cost function from $\mathbf{X}_{new}$ is better (that is lower) than the cost function from $\mathbf{X}_i$, current $\mathbf{X}_i$ is replaced in population P by $\mathbf{X}_{new}$.

Next i.

Step 3.   Select the parameter vector $\mathbf{X}_{best}$ (best fuzzy model parameter set) with the lowest cost (= MSE) function from population P. Extract from $\mathbf{X}_{best}$ all parameters for defining adjusted type-2 membership functions in the rules, i.e. the parameters LL, LR, ML, MR, RL, RR for all input terms and the parameters L, ML, MR, R for all outputs terms.

Step 4.   Terminate the algorithm.

Fig. 3.33 The DE-based parameter optimization used for training of the T2FLNN

1994), an effective tool can be created for approximating any nonlinear function (Buckley and Hayashi 1994).

It can be shown that type-2 fuzzy neural networks are universal approximators as well. The proof, based on the Stone-Weierstrass theorem, for an interval type-2 fuzzy neural network (IT2FNN) can be found in (Castillo et al. 2013).

Theorem 3.3 *Stone-Weierstrass Theorem* (Castillo et al. 2013). Let Z be a set of real continuous functions on a compact set U. If:

1. Z is an algebra (i.e. Z is closed under addition, multiplication, and scalar multiplication)
2. Z separates points on U, i.e. for $\forall \mathbf{x}, \mathbf{y} \in U, \mathbf{x} \neq \mathbf{y}$, there exists $f \in Z$ such that $f(\mathbf{x}) \neq f(\mathbf{y})$
3. Z vanishes at no point of U, i.e. $\forall \mathbf{x} \in U$ there exists $f \in Z$ such that $f(\mathbf{x}) \neq 0$,

then the uniform closure of Z consists of all real continuous functions on U, i.e. (Z, d_∞) is dense in $(C[U]. d_\infty)$ (Scarborough and Stone 1966; Edwards 1995; Rudin 1976)

Theorem 3.4 *Universal approximation theorem* (Castillo et al. 2013). For any given real continuous function $g(\mathbf{u})$ on compact set $U \subset R^n$ and arbitrary $\varepsilon > 0$ there exists $f \in Y$ such that $\sup_{\mathbf{x} \in U}(|g(\mathbf{x}) - f(\mathbf{x})|) < \varepsilon$.

References

Abiyev R (2009), Fuzzy wavelet neural network for prediction of electricity consumption, Artificial Intelligence for Engineering Design, Analysis and Manufacturing 23, 109–118.

Abiyev R, Kaynak O (2008) Fuzzy wavelet neural networks for identification and control of dynamic plants—a novel structure and a comparative study. IEEE Transactions on Industrial Electronics 55, 3133–3140.

Abiyev R, Kaynak O, Kayacan E (2013) A type-2 fuzzy wavelet neural network for system identification and control. Journal of the Franklin Institute 350. 1658–1685.

Aliev RA, Aliev RR (2001b) Soft computing and its applications. World Scientific, 2001, 465 p.

Aliev RA, Fazlollahi B, Vahidov R (2001) Genetic algorithm-based learning of fuzzy neural networks. Part 1: feed-forward fuzzy neural networks. Fuzzy Sets and Systems. v118. 351–358.

Aliev RA, Fazlollahi B, Aliev RR (2004) Soft Computing and its Applications in Business and Economics. Springer, Series: Studies in Fuzziness and Soft Computing, Vol. 157, 2004, XVI, 446 p.

Aliev RA, Fazlollahi B, Guirimov BG, Aliev RR (2008) Recurrent Fuzzy Neural Networks and Their Performance Analysis. In: Hu X and Balasubramaniam P (eds) Recurrent Neural Networks.

Aliev RA, Guirimov BG, Fazlollahi B, Aliev RR (2009) Evolutionary algorithm-based learning of fuzzy neural networks. Part 2: Recurrent fuzzy neural networks. Fuzzy Sets and Systems archive, Volume 160 Issue 17, 2553–2566.

Aliev RA, Pedrycz W, Guirimov B, Aliev RR, Ilhan U, Babagil M, Mammadli S (2011) Type-2 fuzzy neural networks with fuzzy clustering and differential evolution optimization. Information Sciences, Volume 181 Issue 9, 1591–1608.

Al-Rousana M, Assaleh K (2011) A wavelet-and neural network-based voice system for a smart wheel chair control. Journal of the Franklin Institute 348 (1) 90–100.

Angelov P (2004) A fuzzy controller with evolving structure, Information Sciences 161 (1–2)
    21–35
Buckley JJ (1992) Universal fuzzy controllers. Automatica, vol. 28, 6, 1245–1248.
Buckley JJ, Hayashi Y (1994) Can fuzzy neural nets approximate continuous fuzzy functions?.
    Fuzzy Sets and Systems, vol. 61, 1, 43–51, 1994.
Bullinaria J (2013) Course Material, Introduction to Neural Networks, http://www.cs.bham.ac.uk/
    ~jxb/inn.html
Casillas J et al. (eds) (2003) Interpretability issues in fuzzy modeling, Springer, Berlin
Castillo O, Castro JR, Melin P, Rodriguez-Diaz A (2013) Universal approximation of a class of
    interval type-2 fuzzy neural networks in nonlinear identification. Advances in Fuzzy Sys-tems.
    Volume 2013, Article ID 136214.
Castro JR, Castillo O, Melin P, Rodriguez-Diaz A, Mendoza O (2009) Universal approximation of
    a class of interval type-2 fuzzy neural networks illustrated with the case of non-linear
    identification. In: Proceedings of IFSA-EUSFLAT 2009.
Ciaramella A, Tagliaferri R, Pedrycz W (2005) The genetic development of ordinal sums, Fuzzy
    Sets and Systems 151 (2) 303–325
Ciaramella A, Tagliaferri R, Pedrycz W, Di Nola A (2006), Fuzzy relational neural network,
    International Journal of Approximate Reasoning 41 (2), 146–163
Dickerson J, Lan M (1995) Fuzzy rule extraction from numerical data for function approximation,
    IEEE Transactions on System, Man, and Cybernetics—Part B 26, 119–129
Edwards R (1995) Functional analysis: theory and applications, Dover, 1995
Fuller R (2000) Introduction to neuro-fuzzy systems. Springer, 289 p.
Gobi A, Pedrycz W (2006) The potential of fuzzy neural networks in the realization of approx-
    imate reasoning engines, Fuzzy Sets and Systems 157 (22) 2954–2973
Golden R (1996) Mathematical methods for neural network analysis and design, MIT Press,
    Cambridge, MA
Ham FM, Kostanic I (2001) Principles of neurocomputing for science and engineering. McGraw
    Hill.
Haykin S (1999) Neural networks: a comprehensive foundation. Prentice Hall
Hidalgo D, Castillo O, Melin P (2009) Type-1 and type-2 fuzzy inference systems as integration
    methods in modular neural networks for multimodal biometry and its optimization with genetic
    algorithms. Information Sciences, Volume 179, Issue 13, 2123–2145.
Hirota K, Pedrycz W (1994) OR/AND neuron in modeling fuzzy set connectives, IEEE Trans-
    actions on Fuzzy Systems 2, 151–161.
Hirota K, Pedrycz W (1999) Fuzzy relational compression, IEEE Transactions on Systems, Man,
    and Cybernetics—Part B29, 407–415.
Ho D, Zhang P-A, Xu J (2001) Fuzzy wavelet networks for function learning. IEEE Transactions
    on Fuzzy Systems 9 (1) 200–211.
Hornik K, Stinchcombe M, White H (1989) Multilayer feedforward networks are universal
    approximators. Neural Networks, vol. 2, 5, 359–366.
Huang H, Wu C (2009) Approximation capabilities of multilayer fuzzy neural networks on the set
    of fuzzy-valued functions. Information Sciences. Volume 179, Issue 16, 2762–2773.
Ishibuchi H, Kwon K, Tanaka H (1995) A learning algorithm of fuzzy neural networks with
    triangular fuzzy weights, Fuzzy Sets and Systems, 71, 277–293.
Jang J, Sun C, Mizutani E (1997), Neuro-fuzzy and soft computing. Prentice-Hall, Upper Saddle
    River, NJ
Karnik NN and Mendel JM (1999a) Type-2 fuzzy logic systems. IEEE Trans. Fuzzy Syst., Vol.
    7, 643–658.
Kosko B (1991) Neural networks and fuzzy systems. Prentice-Hall, Englewood Cliffs, NJ
Kugarajah T, Zhang Q (1995) Multidimensional wavelet frames. IEEE Transactions on Neural
    Networks 6(6) 1552–1556.
Leung F, Lam H, Ling S, Tam P (2003) Tuning of the structure and parameters of a neural network
    using an improved genetic algorithm. IEEE Transactions on Neural Networks, Vol. 14, No. 1.

Liu P, Li H (2004) Fuzzy neural network theory and application. World Scientific, 5 Toh Tuck Link, Singapore 596224.

Mamdani EH (1977) Application of fuzzy logic to approximate reasoning using linguistic synthesis, IEEE Trans. Computers, 26(12):1182–1191.

Mamdani EH, Assilian S (1975) An experiment in linguistic synthesis with a fuzzy logic controller, International Journal of Man–machine Studies, 7(1):1–13.

Markowska-Kaczmar U, Trelak W (2005), Fuzzy logic and evolutionary algorithm—two techniques in rule extraction from neural networks, Neurocomputing 63, 359–379.

Michalewicz Z (1996), Genetic Algorithms + Data Structures = Evolution Programs. Third ed., Springer, Heidelberg

Mitra S, Pal S (1994) Logical operation based fuzzy MLP for classification and rule generation, Neural Networks 7, 353–373

Mitra S, Pal S (1995) Fuzzy multiplayer perceptron inferencing and rule generation, IEEE Transactions on Neural Networks 6, 51–63.

Negnevitsky M (2005) Artificial Intelligence – a guide to intelligent systems. Addison-Wesley, Pearson Education, Second Edition.

Nobuhara H, Hirota K, Sessa S, Pedrycz W (2005) Efficient decomposition methods of fuzzy relation and their application to image decomposition, Applied Soft Computing 5 (4) 399–408.

Nobuhara H, Pedrycz W, Sessa S, Hirota K (2006), A motion compression/reconstruction method based on max t-norm composite fuzzy relational equations, Information Sciences 176 (17) 2526–2552

Pal S, Mitra S (1999), Neuro-fuzzy pattern recognition, Wiley, NewYork

Pedrycz W (1991a), Processing in relational structures: fuzzy relational equations, Fuzzy Sets and Systems 40, 77–106.

Pedrycz W (1991b), Neurocomputations in relational systems, IEEE Transactions on Pattern analysis and Machine Intelligence 13, 289–297.

Pedrycz W (1993), Fuzzy neural networks and neurocomputations, Fuzzy Sets and Systems 56, 1–28.

Pedrycz W (2004) Heterogeneous fuzzy logic networks: fundamentals and development studies, IEEE Transactions on Neural Networks 15, 1466–1481.

Pedrycz W (2007) Genetic tolerance fuzzy neural networks: from data to fuzzy hyperboxes, Neurocomputing 70 (7–9) 1403–1413

Pedrycz W, Gomide F (1998) An introduction to fuzzy sets: Analysis and Design, MIT Press, Cambridge, MA

Pedrycz W, Gomide F (2007) Fuzzy systems engineering: Toward human-centric computing, Wiley, Hoboken, NJ

Pedrycz W, Reformat M (2005) Genetically optimized logic models, Fuzzy Sets and Systems 150 (2), 351–371

Pedrycz W, Rocha A (1993), Knowledge-based neural networks, IEEE Transactions on Fuzzy Systems 1 254–266

Pedrycz W, Lam P, Rocha A (1995) Distributed fuzzy modeling, IEEE Transactions on Systems, Man and Cybernetics—Part B5, 769–780.

Price K, Storn R, Lampinen J (2005) Differential evolution – a practical approach to global optimization. Springer, Berlin.

Rudin W (1976) Principles of mathematical analysis, McGraw-Hill, New York, NY, USA

Scarborough C, Stone A (1966) Products of nearly compact spaces, Transactions of the American Mathematical Society, vol. 124, 131–147

Setnes M, Babuska R, Vebruggen H (1998) Rule-based modeling: precision and transparency, IEEE Transactions on System, Man, and Cybernetics—Part C28, 165–169.

Sharifi A, Shoorehdeli MA, Teshnehlab M (2012) Identification of cement rotary kiln using hierarchical wavelet fuzzy inference system. Journal of the Franklin Institute 349 (1) 162–183.

Takagi T and Sugeno M (1983) Derivation of fuzzy control rules from human operators control actions. In: Proc. IFAC Symp. on Fuzzy Information, Knowledge Representation and Decision Analysis, 55–60.

Thuillard M (2001) Wavelets in soft computing, World Scientific Press.

Wang L-X, Mendel JM (1992a) Fuzzy basis functions, universal approximation, and orthogonal least-squares learning. IEEE Transactions on Neural Networks, vol. 3, 5, 807–814.

Wang L-X, Mendel JM (1992b) Generating fuzzy rules by learning from examples. IEEE Transactions on Systems, Man and Cybernetics, vol. 22, 6, 1414–1427.

Wang CH, Cheng CS, Lee TT (2004) Dynamical optimal training for interval type-2 fuzzy neural network (T2FNN). IEEE Trans. on Systems, Man, Cybernetics Part-B, 34(3):1462–1477

Yager R (2001), Uninorms in fuzzy systems modeling, Fuzzy Sets and Systems 122, 167–175

Yager R, Rybalov A (1996) Uninorm aggregation operators, Fuzzy Sets and Systems 80 111–122

Zadeh LA (1965) Fuzzy sets. Information and Control, vol. 8, 3, 338–353.

Zhang Q, Benveniste A (1992) Wavelet networks. IEEE Transactions on Neural Networks 3 (6) 889–898.

Zhang J, Walter G, Miao Y, Lee WNW (1995) Wavelet neural networks for function learning, IEEE Transactions on Signal Processing 43(6) 1485–1497.

Chapter 4
Type-2 Fuzzy Clustering

4.1 Introductory Notes: Neural Networks and Clustering

As was pointed out in Chap. 3 of this book, the required design for a network (in terms of quantity of hidden layers and hidden neurons) to perform its operation accurately is very dependent of the complexity of the function or classification to be learned. To get a good network it is important to provide efficient training. Among the other factors, the efficiency of training depends on the features of training data: amount of noise and inconsistencies in data, the number and distribution of data, data ranges etc.

It is proven by a number of researchers (Bullinaria 2013; Negnevitsky 2005; Dickerson and Lan 1995; Huang and Wu 2009) that preprocessing of data before applying them as training data patterns for neural networks can significantly improve the efficiency of training as well as the quality of resultant network system. Too few and too many data patterns can result in poor generalization ability, too many data can lead also to long training time. Inconsistent data can prevent the error function during the training to converge to a required level at all. Figure 4.1 shows the original function and outputs of the networks trained with different data sets – one set with too many noisy data and the other one with lack of data.

To train the networks in presence of noisy data many approaches are used. The appropriate techniques include adding a regularization term into the training error function, separating a validation data set, randomly picking data from the data set during the training, adding noise to the data, and finally, data clustering. Clustering can part of training strategy for such neural networks as RBF and logical and ordinary fuzzy neural networks in case of presence of large sets of crisp and noisy training data. Crisp data are filtered and fuzzy training sets are formed as result of a fuzzy clustering method. Let's consider in some detail, the standard Fuzzy C-Means clustering algorithm (Bezdek 1973, 1981; Rhee 2007).

© Springer International Publishing Switzerland 2014   153
R.A. Aliev, B.G. Guirimov, *Type-2 Fuzzy Neural Networks and Their Applications*,
DOI 10.1007/978-3-319-09072-6_4

Fig. 4.1 Cases of outputs of neural networks trained on different data sets

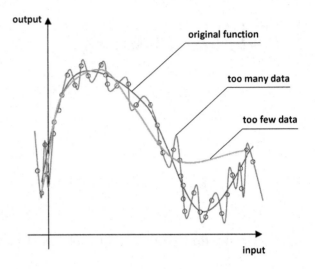

Fuzzy C-Means Algorithm The Fuzzy C-Means (FCM) is an algorithm to partition numerical data vectors into a nonempty set of clusters (Bezdek 1973, 1981; Valente de Oliveira and Pedrycz 2007).

Let's consider a data set consisting of n vectors of dimension D:

$$X = \{\mathbf{x}_1, \mathbf{x}_2, \ldots, \mathbf{x}_n\}, X \subseteq \mathfrak{R}^D.$$

In FCM every considered data vector maintains a fuzzy membership grade to each of the produced classes $\mathbf{v}_j$, $j = 1, \ldots, c$:

$$u_{ij} \equiv u_j(\mathbf{x}_i) \in [0, 1],$$
$$\sum_{j=1}^{c} u_{ij} = 1, \quad i = 1, \ldots, n, \quad j = 1, \ldots, c. \tag{4.1}$$

Where c is the number of produced clusters and u_{ij} is the membership grade of the i-th data vector ($\mathbf{x}_i$) to j-th cluster $\mathbf{v}_j$.

The positions of clusters are computed as:

$$\mathbf{v}_j = \frac{\sum_{i=1}^{n} \mathbf{x}_i u_{ij}^m}{\sum_{i=1}^{n} u_{ij}^m} \tag{4.2}$$

where m is the fuzzification parameter (often called a fuzzifier) and controls the fuzziness of the produced clusters. The dependence of fuzzy cluster boundaries of the fuzzifier m is illustrated in Fig. 4.2.

Fig. 4.2 Effect of the fuzzifier m on cluster boundaries

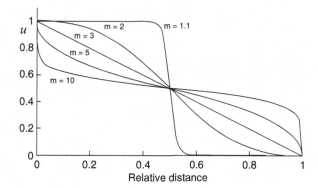

The optimal membership degrees u_{ji} are found by minimization of the following objective function:

$$J\left(u_{ij}\right) = \sum_{j=1}^{c} \sum_{i=1}^{n} u_{ij}^{m} d_{ij}^{2}, \qquad (4.3)$$

subject to the constraints (4.1),

where $d_{ij} \equiv d(\mathbf{x}_i, \mathbf{v}_j)$ is the chosen vector distance norm, e.g. Euclidean.

The FCM clustering algorithm starts from randomly initializing of all cluster positions and continually applying the following formula to update the memberships and, accordingly, the vector positions (4.2):

$$u_{ij} = \frac{1}{\sum_{l=1}^{c} \left(d_{ij}/d_{il}\right)^{2/(m-1)}} \qquad (4.4)$$

The algorithm is stopped once the process has converged with no further significant changes to the positions.

4.2   Interval and General Type-2 FCM Clustering Method

When dealing with fuzzy clustering, cf. (Hwang and Rhee 2007), there are several essential parameters whose values need to be decided upon in advance. We can think of uncertainty, which is inherently associated with the selection of the specific numeric values of these parameters. In the FCM-like family of fuzzy clustering, the fuzzification coefficient (fuzzifier) plays a visible role as it directly translates into the shape (geometry) of resulting fuzzy clusters. Let us recall that for the values of m close to 1, the membership functions become very close to the characteristic functions of sets whereas higher values of m (say, over 3 or 4) result in "spiky" membership functions (refer to Fig. 4.2 above).

The IT2 FCM method (Linda and Manic 2012; Hwang and Rhee 2007) considers an interval valued fuzzifier $[m_L, m_R]$ instead a crisp value as in the traditional FCM. The interval memberships $\left[\underline{u}_{ij}, \overline{u}_{ij}\right]$ of vector $\mathbf{x}_i$ to cluster $\mathbf{v}_i$ can be computed as follows:

$$
\underline{u}_{ij} = \min \left(\frac{1}{\sum\limits_{l=1}^{c} \left(d_{ij}/d_{il}\right)^{2/(m_L-1)}}, \frac{1}{\sum\limits_{l=1}^{c} \left(d_{ij}/d_{il}\right)^{2/(m_R-1)}} \right),
$$

$$
\overline{u}_{ij} = \max \left(\frac{1}{\sum\limits_{l=1}^{c} \left(d_{ij}/d_{il}\right)^{2/(m_L-1)}}, \frac{1}{\sum\limits_{l=1}^{c} \left(d_{ij}/d_{il}\right)^{2/(m_R-1)}} \right).
$$

$$(4.5)$$

The interval cluster positions can be expressed as follows (Linda and Manic 2012; Hwang and Rhee 2007):

$$
\widetilde{\mathbf{v}}_j = \left[v_j^L, v_j^R \right] = \sum_{u_{1j}} \cdots \sum_{u_{nj}} 1 \left/ \frac{\sum\limits_{i=1}^{n} x_i u_{ij}^m}{\sum\limits_{i=1}^{n} u_{ij}^m} \right.
$$

$$(4.6)$$

where m switches from m_L to m_R as in Eq. 4.5. The values for the left and right cluster boundaries in each dimension can be computed by sorting the patterns in particular dimension and then applying the Karnik-Mendel iterative procedure (Karnik and Mendel 2001).

The value of m has a direct impact on the location and quality of the cluster partition. However, it is difficult to express the notion of fuzziness in the input data using precise real value for T1 FCM algorithm or even as an interval value for IT2 FCM algorithm (Bezdek 1973, 1981; Valente de Oliveira and Pedrycz 2007; Rhee 2007). To alleviate this issue, the proposed in (Linda and Manic 2012) General Type-2 (GT2) FCM algorithm allows linguistic expression of the fuzzifier value using terms such as "small" or "high" modeled as T1 FSs (Fig. 4.3). The resulting cluster membership functions can be implemented as GT2 FSs represented using the α-planes theorem (Liu 2008; Mendel et al. 2009), considered below.

Fig. 4.3 Linguistic representation of the fuzzifier m

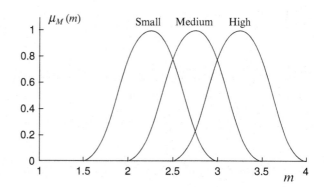

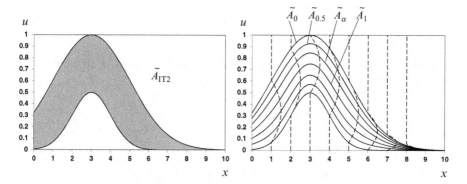

Fig. 4.4 Different representations of T2 FS – IT2 and α-plane based

An α-plane $\widetilde{A}_\alpha$ of a GT2 FS $\widetilde{A}$ can be defined as the union of all primary memberships of $\widetilde{A}$ with secondary grades greater than or equal to α (Liu 2008; Tahayori et al. 2006):

$$\widetilde{A}_\alpha = \int_{\forall x \in X} \int_{\forall u \in J_x} \left\{ (x, u) \big| f_x(u) \geq \alpha \right\}. \tag{4.7}$$

Then:

Theorem *Liu's representation theorem* (Liu 2008). GT2 FS $\widetilde{A}$ can be constructed as a composition of all of its individual α-level T2 FSs as follows:

$$\widetilde{A} = \bigcup_{\alpha \in [0,1]} \alpha / \widetilde{A}_\alpha. \tag{4.8}$$

Figure 4.4 (Linda and Manic 2012) demonstrates an example of T2 FS represented as an interval T2 FS and as union of its α-planes.

Then, the linguistic fuzzifier $\tilde{m}$ can be represented using its a-cuts as:

$$\tilde{m} = \cup \alpha / \left[m^L(\alpha), m^R(\alpha) \right]$$

The GT2 fuzzy membership function for cluster $\mathbf{v}_j$ can be found by combining T1 fuzzy memberships (of data vector $\mathbf{x}_i$ to cluster $\mathbf{v}_j$) from all data vectors:

$$\tilde{u}_j = \sum_{x_i \in X} \tilde{u}_j(\mathbf{x}_i). \qquad (4.9)$$

Here $\tilde{u}_j(x_i)$ can be interpreted as secondary MFs of the GT2 cluster MF $\tilde{u}_j$ sampled at the locations of vectors $\mathbf{x}_i$.

The secondary MF $\tilde{u}_j(\mathbf{x}_i)$ can be expressed by its α-cuts as follows:

$$\tilde{u}_j(\mathbf{x}_i) = \underset{\alpha \in [0,1]}{\cup} \alpha / \left[u_j^L\left(\mathbf{x}_i | \alpha\right), u_j^R\left(\mathbf{x}_i | \alpha\right) \right]. \qquad (4.10)$$

Then an α-plane of the GT2 cluster MF is produced by combining all α-cuts over all data vectors:

$$u_j(\alpha) = \sum_{\mathbf{x}_i \in X} \left[u_j^L\left(\mathbf{x}_i | \alpha\right), u_j^R\left(\mathbf{x}_i | \alpha\right) \right],$$

$$u_j^L\left(\mathbf{x}_i | \alpha\right) = \min \left(\frac{1}{\sum\limits_{l=1}^{c} \left(d_{ij}/d_{il}\right)^{2/(m^L(\alpha)-1)}}, \frac{1}{\sum\limits_{l=1}^{c} \left(d_{ij}/d_{il}\right)^{2/(m^R(\alpha)-1)}} \right),$$

$$u_j^R\left(\mathbf{x}_i | \alpha\right) = \max \left(\frac{1}{\sum\limits_{l=1}^{c} \left(d_{ij}/d_{il}\right)^{2/(m^L(\alpha)-1)}}, \frac{1}{\sum\limits_{l=1}^{c} \left(d_{ij}/d_{il}\right)^{2/(m^R(\alpha)-1)}} \right) \qquad (4.11)$$

The computation of fuzzy cluster positions is done as follows:

$$\tilde{\mathbf{v}}_j = \sum_{u_{1j}} \cdots \sum_{u_{nj}} f\left(\tilde{u}_{1j}\right) * \ldots * f\left(\tilde{u}_{nj}\right) \Bigg/ \frac{\sum\limits_{i=1}^{n} x_i \tilde{u}_{ij}^m}{\sum\limits_{i=1}^{n} \tilde{u}_{ij}^m}$$

$$= \underset{\alpha \in [0,1]}{\cup} \alpha / \left[\mathbf{v}_j^L(\alpha), \mathbf{v}_j^R(\alpha) \right]. \qquad (4.12)$$

where * denotes a selected T-norm operation.

The computation associated with the above formula (4.12) can be done using the well-known EKM algorithm (Liu 2008; Mendel et al. 2009)

4.3 Interval Type-2 Fuzzy Clustering Using DE

The clustering algorithms built upon gradient-based optimization (including the FCM itself) exhibit some disadvantages (Aliev et al. 2011). One of significant drawbacks is that they may not produce a global minimum but instead could get stuck in some local minimum. On the other hand, the standard iterative scheme may not be applicable directly to the considered problem, especially when various distance functions other than the Euclidean one are being used. As pointed out in (Pedrycz and Hirota 2007), a better approach could be to consider a population-based algorithm such as genetic algorithm, DE, or PSO.

Based on the experimental work reported in (Ozkan and Turksen 2007) we reformulate the clustering problem (4.3) as:

$$J_{m1} = \sum_{i=1}^{n}\sum_{j=1}^{c} u_{ij}^{m_1} \left\| \mathbf{x}_i - \mathbf{v}_j^{(1)} \right\|^2 \rightarrow \min,$$

$$J_{m2} = \sum_{i=1}^{n}\sum_{j=1}^{c} u_{ij}^{m_2} \left\| \mathbf{x}_i - \mathbf{v}_j^{(2)} \right\|^2 \rightarrow \min,$$

$$u_{ij} = \frac{1}{\sum_{l=1}^{c}\left(d_{ij}/d_{il}\right)^{2/(m-1)}}$$

(4.13)

subject to constraints:

$$0 < \sum_{i=1}^{n} u_{ij} < n \; (j = 1, \; 2, \; \ldots, \; c) \text{ and } \sum_{j=1}^{c} u_{ij} = 1 \; (i = 1, \; 2, \; \ldots, \; n).$$

The cluster vector $\widetilde{\mathbf{v}}_i$ is formed as:

$$\widetilde{\mathbf{v}}_i = \left[\min\left(\mathbf{v}_i^{(1)}, \mathbf{v}_{Ind_i}^{(2)} \right), \max\left(\mathbf{v}_i^{(1)}, \mathbf{v}_{Ind_i}^{(2)} \right) \right],$$

(4.14)

where $Ind_i = \underset{j}{\arg\min} \left\| \mathbf{v}_i^{(1)}, \mathbf{v}_j^{(2)} \right\|$.

Here the fuzzifier $\widetilde{m}$ represents interval (type-1 fuzzy) value ($[m_1, m_2]$); $\widetilde{\mathbf{v}}_j$ is the prototype of the j-th cluster generated by fuzzy clustering; $\widetilde{u}_{ij}$ is the membership degree of the i-th data belonging to the j-th cluster represented by the data vector $\widetilde{\mathbf{v}}_j$.

The above conclusion is based on the result that the change in the location of the prototypes associated with a change of the values of m is of monotonic character. The above assumption allows us to replace Eq. 4.3 by Eq. 4.13 and significantly reduce the computational burden when searching for cluster centers using different optimization methods. As shown in (Ozkan and Turksen 2007) the meaningful range for m for most problems is [1.4, 2.6].

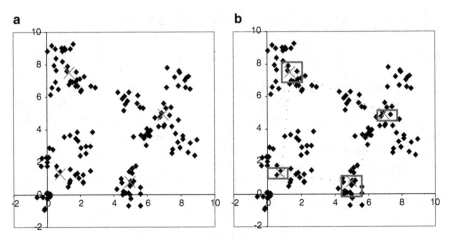

Fig. 4.5 Clusters produced by: (**a**) the FCM and (**b**) the DE-based type-2 fuzzy clustering

The choice of the number of clusters can be realized on a basis of some cluster validity criterion, like the one suggested in (Chen and Linkens 2004):

$$V(u, c) = \frac{1}{n}\sum_{k=1}^{n} \max\left(u_{ij}\right) - \frac{1}{K}\sum_{i=1}^{c-1} \sum_{j=i+1}^{c} \underbrace{\left[\frac{1}{n}\sum_{k=1}^{n} \min\left(u_{ik}, u_{jk}\right)\right]}_{i}, \quad \text{where } K = \sum_{i=1}^{c-1} i.$$

(4.15)

The number of clusters can be defined on the basis of the maximal value of V, which is picked up for different values of $\widetilde{m}$. The minimum number of clusters and, therefore, the number of the fuzzy rules themselves is very important for interpretability reasons. However, this number may need to be increased based on the required accuracy (expressed via the MSE performance index).

Clustering is completed through the minimization of the values of the objective functions (4.13) using an evolutionary algorithm. Here we suggest using DE – a simple, yet powerful global optimizer.

An illustrative example of the clustering problem is shown in Fig. 4.5 (Aliev et al. 2011). There is a collection of two-dimensional data with clearly visible four clusters. The data points are separated into four clusters. Figure 4.5a shows the clusters found by the standard FCM with $m = 2$. Figure 4.5b shows interval type-2 fuzzy clustering result for the same problem. In Fig. 4.5b the squares around the cluster centers, shown by cross-signs (×), incorporate the cluster centers found for a variety of m within the range [1.05, 9.0]; the clusters found with $m = 2$ are connected with the thin dotted line. It should be noted that, following our experiments, DE converges more successfully than the standard FCM within the considered range of m. It can be seen from Fig. 4.5a, b that in the case when $m_1 = m_2$, i.e. for type-1 FCM the cluster centers are located within the bounds of the cluster

Table 4.1 Example cluster centers

J	1	2	3	4	5	6	7	8	9	10
$C_1=$	1.0	2.0	3.0	4.0	5.0	4.0	3.0	2.0	1.0	2.0
$C_2=$	2.0	7.0	8.0	3.0	7.0	8.0	4.0	7.0	8.0	5.0
$C_3=$	1.0	1.0	1.5	1.0	1.0	1.5	1.0	1.0	1.0	1.0
$C_4=$	4.0	6.0	7.0	5.0	8.0	3.0	5.0	4.0	9.0	9.0
$C_5=$	2.0	7.5	8.0	3.0	7.5	8.0	4.0	7.0	8.0	4.5
$C_6=$	1.0	1.0	1.0	1.0	1.0	2.0	1.0	1.0	1.0	1.0
$C_7=$	1.0	6.0	3.0	3.0	5.0	5.0	2.0	6.0	6.0	8.0

Table 4.2 FCM performance

Experiment	Objective function ($J_{m=2.5}$)
1	88.148
2	96.902
3	84.988
4	94.693
5	118.788
6	84.988

squares obtained by type-2 FCM. At the same time, the cluster squares allows capturing more uncertainty in data than one-point cluster centers.

Even for narrower ranges of m, say $m=[1.4, 2.6]$ DE as a global search algorithm is expected to be more advantageous than the standard FCM for the case of large number of highly-dimensional data vectors. To show this, we run the following experiment. Consider seven vectors $C_1, C_2, C_3, C_4, C_5, C_6, C_7$ of dimension 10 with locations shown in Table 4.1.

We generate 1,000 data vectors of dimension 10 on the basis of 7 original vectors $C_1, C_2, C_3, C_4, C_5, C_6, C_7$ as follows:

$$p_{ik} = C_{((i-1)\%7+1),k} + 1.5 \cdot (rand(0, 1) - 0.5); \quad i = 1, 2, \ldots, 1000;$$
$$k = 1, 2, \ldots, 10,$$

where $rand(0,1)$ is a random number drawn from the uniform distribution $[0,1]$ and % is the operation of determining a remainder of integer division.

Consecutive five runs of the standard FCM algorithm applied to the generated vectors for finding seven cluster centers with $m=2.5$ after 1,000 iterations (at which point the FCM has converged and no more iterations are required) have ended up with the values of the objective function (4.13) with Euclidean distance function shown in Table 4.2 (Aliev et al. 2011).

Table 4.3 DE clustering result

J	1	2	3	4	5	6	7	8	9	10
v_1	1	1.95	2.98	4.01	4.97	3.98	3	2.01	1	1.98
v_2	1.98	7.05	7.98	2.99	6.99	7.97	4	6.96	7.94	5.01
v_3	1.99	7.46	8.02	2.99	7.48	7.96	4	6.96	8.06	4.5
v_4	3.99	5.97	6.99	4.95	8.01	3	4.95	3.99	8.98	8.98
v_5	0.99	5.99	3.01	2.99	4.99	4.99	2.01	5.99	6	8
v_6	0.99	0.99	1.49	0.99	0.99	1.5	1	0.99	1.01	1
v_7	1	0.99	1.01	0.99	0.99	1.98	0.99	0.98	0.99	0.99

The application of the global search DE algorithm ($ps = 200$) allowed us to minimize the objective function (4.3 or 4.13) down to $J_{m=2.5} = 79.076$. The clusters found by the DE based clustering are shown in Table 4.3 (Aliev et al. 2011):

Note that the actual minimum value of the objective function for this example (found by the optimization with initial cluster centers set to the original seven data vectors instead of random numbers) is approximately 77.7.

The parameters to be optimized by the DEO when minimizing the objective function are the centers of the clusters, $v_i^{(1)}$, $v_i^{(2)}$, $i = 1, 2, \ldots, c$. The DEO-based clustering algorithm can be formally described as demonstrated in Fig. 4.6.

4.4   Type-2 Fuzzy Neural Network with Clustering

Clustering can be a very useful technique to improve the performance of training of type-2 neural networks (Aliev et al. 2011). On availability of large amounts of possibly noisy raw input-to-output data, fuzzy type-2 clustering can be used for preparing valid training data set and as an initial part of training. Clustering allows generation of initial type-2 fuzzy If-Then rule base from the data by providing optimal numbers of fuzzy type-2 terms for input and output variables and values for parameters defining starting membership functions for the terms.

Once the initial rule base is obtained, its parameters are further adjusted by applying an evolutionary optimization algorithm for a considered fuzzy type-2 inference system, modeled by a specific type-2 fuzzy neural network. Thus, the training process consists of two phases: (1) generation of initial rule-base using a type-2 fuzzy clustering method and (2) further adjusting the rule base for a particular type-2 fuzzy logical inference system model.

To be specific, without limiting generality, let's consider the type-2 fuzzy neural network described in Sect. 3.6 (Fig. 3.29).

Given n input-output pairs of data $\{X_1/Y_1, X_2/Y_2, \ldots, X_n/Y_n\}$ and the required accuracy of the model ($\varepsilon \geq 0$), we form the minimal number of the rules and

DEO_Based_Clustering(data vectors)

1. Set the number of clusters $c = 2$ (start with the minimal number of clusters, that is $c = 2$);

2. Compute cluster centers for m_1 : $\mathbf{v}_i^{(1)}$ ($i = 1, 2, ..., c$)

   2.1. Create a population of random solutions (population size is $ps=c*10$) for cluster centers $\mathbf{v}_i^{(1)}$: $P = \{\mathbf{S}_1, \mathbf{S}_2, ..., \mathbf{S}_{ps}\}$ (Each $\mathbf{S}_j$ ($j = 1, 2, ..., ps$) represent a combination of cluster centers for given data).

   2.2. Prepare the standard DE strategy with $PopSize=ps$, , $f=0.9$, $cr=1$ and population P. Set the *Cost* function to J_{m1}.

   2.3. Execute the DE strategy until a maximum number of iteration reached or no progress in optimization of the cost function detected.

   2.4. Retrieve optimal cluster centers $\mathbf{v}_i^{(1)}$ from $\mathbf{S}_{best}$, where $Cost(\mathbf{S}_{best}) \leq Cost(\mathbf{S}_r)$, $r = 1, 2, ..., ps$.

3. Compute cluster centers for m_2 : $\mathbf{v}_i^{(2)}$ ($i = 1, 2, ..., c$)

   3.1. Create the population of random solutions (population size is $ps=c*10$) for cluster centers $\mathbf{v}_i^{(2)}$: $P = \{\mathbf{S}_1, \mathbf{S}_2, ..., \mathbf{S}_{ps}\}$.

   3.2. Prepare the standard DE strategy with $PopSize=ps$, , $f=0.9$, $cr=1$ and population P. Set the *Cost* function to J_{m2}.

   3.3. Execute the DE strategy until a maximum number of iteration reached or no progress in optimization of the cost function detected.

   3.4. Retrieve optimal cluster centers $\mathbf{v}_i^{(2)}$ from $\mathbf{S}_{best}$, where $Cost(\mathbf{S}_{best}) \leq Cost(\mathbf{S}_r)$, $r = 1, 2, ..., ps$.

4. Check if the validity criterion (4.15): $V = \min(V_{m1}, V_{m2})$ attains higher value than a predefined threshold (e.g., 0.6). If yes, then go to step 5, otherwise increase c by one ($c = c+1$) and repeat from step 2.

5. On the basis of optimal cluster centers $\mathbf{v}_i^{(1)}$ and $\mathbf{v}_i^{(2)}$ form the interval cluster centers:

   $\tilde{\mathbf{v}}_i = [\min(\mathbf{v}_i^{(1)}, \mathbf{v}_{Ind_i}^{(2)}), \max(\mathbf{v}_i^{(1)}, \mathbf{v}_{Ind_i}^{(2)})]$, where $Ind_i = \arg\min_j \left\| \mathbf{v}_i^{(1)}, \mathbf{v}_j^{(2)} \right\|$

6. Stop

Fig. 4.6 DEO based clustering algorithm

parameters of the type-2 membership functions (for instance, six parameters describing each interval valued triangular membership function described here as $T(LL, LR, ML, MR, RL, RR)$ for input and output terms) so that the error function denoted as (Aliev et al. 2011)

$$Err = \sum_{i=1,n} \|y(\mathbf{X}_i) - Y_i\|,$$

where $y(\mathbf{X})$ is the inference system's numeric output for any given input data vector $\mathbf{X}$ of dimension s: $\mathbf{X} = [x_1 x_2 \ldots x_s]^T)$ satisfies the inequality $Err < \varepsilon$. The fuzzy model output will be obtained on the basis of the inferencing from the following IF-THEN rules:

$$R^i : \text{ IF } x_1 \text{ is } \widetilde{A}_1^i \text{ and } x_2 \text{ is } \widetilde{A}_2^i \text{ and } \ldots x_s \text{ is } \widetilde{A}_s^i \text{ THEN } y \text{ is } \widetilde{B}^i, \qquad (4.16)$$

where x_j ($j = 1, \ldots, s$) and y are input and output variables, respectively; $\widetilde{A}_j^i = 1$, 2, $\ldots$, s) and $\widetilde{B}^i$ are antecedent and consequent type-2 fuzzy sets, respectively. Refer to Sect. 3.6 for details of implementation of the inference.

Assume that we have obtained the cluster centers by a type-2 fuzzy clustering method (Sects. 4.2 and 4.3) $[\mathbf{v}_{1i}, \mathbf{v}_{2i}]$, $i = 1, \ldots, n$. Note that the dimension of the cluster vectors is $(s + 1)$ – with first s dimensions for input variables and the last dimension for the single output variable. The data base will have the number of rules being equal to the number of the clusters n. The initial input and output sets in the rules can be set as follows:

$$\begin{aligned}
\widetilde{A}_j^i &= T(\mathbf{v}_{1i}[j], \mathbf{v}_{1i}[j], \mathbf{v}_{1i}[j], \mathbf{v}_{2i}[j], \mathbf{v}_{2i}[j], \mathbf{v}_{2i}[j]), \quad j = 1, \ldots, s \\
\widetilde{B}^i &= T(\mathbf{v}_{1i}[s+1], \mathbf{v}_{1i}[s+1], \mathbf{v}_{2i}[s+1], \mathbf{v}_{2i}[s+1])
\end{aligned} \qquad (4.17)$$

where $\mathbf{v}_{1i}[j]$ and $\mathbf{v}_{2i}[j]$ are the j-th components of the lower and upper values of the s-dimensional interval-valued cluster vector $\widetilde{\mathbf{v}}_i$, $i = 1, \ldots, n$.

Note that the membership values obtained from clustering can also be used to more properly identify the type-2 fuzzy numbers for parameters $\widetilde{A}_j^i$ and $\widetilde{B}^i$. However, because the actual values are highly dependent on the actual inference implementation, they will better be adjusted at the next stage of training.

Thus, the clustering as the first stage of training of the type-2 fuzzy neural network can be described by the detailed DE-based algorithm presented in Fig. 4.7 (Step 1 of the algorithm presented in Sect. 3.9, Fig. 3.33, is modified in this version to allow for application of the clustering results):

Step 0.   Initialize DE
   Step 0.1.   Construct template parameter vector S of dimension (N_{par}) necessary to accommodate all parameters to be optimized. X consists of the complete set of parameters (LL, LR, ML, MR, RL, or RR) for all input terms and the complete set of parameters (L, ML, MR, or R) for all output terms.
   Step 0.2.   Set the essential parameters of the DE: f (mutation rate), cr (crossover rate), and $PopSize$ (size of population).
   Step 0.3.   Define the cost (objective) function expressed as the Mean Square Error (MSE) between the output of the fuzzy model with the current parameters and desired output (coming from the training data): Cost=

$$E = \frac{1}{n \cdot s_y} \sum_{p=1}^{n} \sum_{i=1}^{s_y} \left(y_{pi}^* - y_{pi} \right)^2 .$$

Step 1.   Generate parameters vectors
   Step 1.1.   Randomly generate $PopSize$ parameter vectors (from respective parameter spaces, e.g., in the range [-1, 1]) and form a population $P = \{\mathbf{X}_1, \mathbf{X}_2, ..., \mathbf{X}_{ps}\}$.
   Step 1.2.   Change the values of one of the population vectors (e.g. $\mathbf{X}_1$) by the data received from type-2 fuzzy clustering method. Import to this population vector the complete parameter sets (LL, LR, ML, MR, RL, or RR) for all input terms and (L, ML, MR, or R) for all output terms.
Step 2.   While the termination condition (say, the number of predefined generations or the required error level) has not been reached, generate new parameter sets:
   Step 2.1.   Choose a next vector $\mathbf{X}_i$ ($i = 1, ..., PopSize$).
   Step 2.2.   Choose randomly three different vectors from P: $\mathbf{X}_{r1}, \mathbf{X}_{r2}, \mathbf{X}_{r3}$ ($1 \le r_1 \neq r_2 \neq r_3 \le PopSize$) each of which is different from $\mathbf{X}_i$.
   Step 2.3.   Generate a trial vector $\mathbf{X}_t = (\mathbf{X}_{r1} - \mathbf{X}_{r2}) f + \mathbf{X}_{r3}$.
   Step 2.4.   Generate a new vector from the trial vector $\mathbf{X}_t$. Individual vector parameters of $\mathbf{X}_t$ are inherited with probability cr and included into the new vector $\mathbf{X}_{new}$. If the cost function from $\mathbf{X}_{new}$ is better (that is lower) than the cost function from $\mathbf{X}_i$, current $\mathbf{X}_i$ is replaced in population P by $\mathbf{X}_{new}$.

   Next i.

Step 3.   Select the parameter vector $\mathbf{X}_{best}$ (best fuzzy model parameter set) with the lowest cost (= MSE) function from population P. Extract from $\mathbf{X}_{best}$ all parameters for defining adjusted type-2 membership functions in the rules, i.e. the parameters LL, LR, ML, MR, RL, RR for all input terms and the parameters L, ML, MR, R for all outputs terms.

Step 4.   Terminate the algorithm.

Fig. 4.7 DE based training of fuzzy type-2 fuzzy neural network with clustering

References

Aliev RA, Pedrycz W, Guirimov B, Aliev RR, Ilhan U, Babagil M, Mammadli S (2011) Type-2 fuzzy neural networks with fuzzy clustering and differential evolution optimization. Information Sciences, Volume 181 Issue 9, 1591–1608.

Bezdek J (1973) Fuzzy mathematics in pattern recognition. Ph.D. dissertation, Appl. Math. Center, Cornell Univ., Ithaca, NY.

Bezdek J (1981) Pattern Recognition with Fuzzy Objective Function Algorithms. New York: Plenum, 1981.

Bullinaria J (2013) Course Material, Introduction to Neural Networks, http://www.cs.bham.ac.uk/~jxb/inn.html

Chen MY, Linkens DA (2004) Rule-base self-generation and simplification for data-driven fuzzy models. J. Fuzzy Sets and Systems 142, 243–265.

Dickerson J, Lan M (1995) Fuzzy rule extraction from numerical data for function approximation, IEEE Transactions on System, Man, and Cybernetics—Part B 26, 119–129

Huang H, Wu C (2009) Approximation capabilities of multilayer fuzzy neural networks on the set of fuzzy-valued functions. Information Sciences, Volume 179, Issue 16, 2762–2773.

Hwang C, Rhee F (2007) Uncertain fuzzy clustering: Interval type-2 fuzzy approach to C-Means, IEEE, 107–120.

Karnik N, Mendel JM (2001) Centroid of a type-2 fuzzy set. Inf. Sci., vol. 132, 195–220.

Linda O, Manic M (2012) General type-2 fuzzy C-means algorithm for uncertain fuzzy clustering. IEEE Transactions on Fuzzy Systems, Vol. 20, No. 5.

Liu F (2008) An efficient centroid type-reduction strategy for general type-2 fuzzy logic sys-tem., Inf. Sci., vol. 178, 2224–2236.

Mendel JM, Liu F, Zhai D (2009) α-plane representation for type-2 fuzzy sets: Theory and applications. IEEE Trans. Fuzzy Syst., vol. 17, no. 5, 1189–1207.

Negnevitsky M (2005) Artificial Intelligence – a guide to intelligent systems. Addison-Wesley, Pearson Education, Second Edition.

Ozkan I, Turksen IB (2007) Upper and Lower Values for the Level of Fuzziness in FCM. Information Sciences, Volume 177, Issue 23, 5143–5152.

Pedrycz W, Hirota K (2007) Fuzzy vector quantization with particle swarm optimization: A study in fuzzy granulation-degranulation information processing. Signal Processing, doi:10.1016/j.sigpro.2007.02.001.

Rhee F (2007) Uncertain fuzzy clustering: Insights and recommendations. IEEE Comput. Intell. Mag., vol. 2, no. 1, pp. 44–56.

Tahayori H, Tettamanzi AG, Antoni GD (2006) Approximated type-2 fuzzy set operations. In: Proc. IEEE Int. Conf. Fuzzy Syst., 2006, 1910–1917.

Valente de Oliveira J, Pedrycz W (eds.) (2007) Advances in Fuzzy Clustering and its Applications. Hoboken, NJ: Wiley.

Chapter 5
Application of Type-2 Fuzzy Neural Networks

5.1 Type-2 Fuzzy Neural Networks for Decision Making

Fuzzy set theory is useful systematic theory when dealing with uncertainty and vagueness in human-originated information. It is due to that fuzzy decision making is more adequate to general human decision behavior and linguistic preferences. The IF-THEN rules can be used for approximate reasoning in decision making. Fuzzy Logic Systems (FLSs) as universal approximators can handle the uncertainties and model the system performance using an easy to understand linguistic labels (such as "low", "medium", "high", etc.) and IF-THEN rules. FLSs processes information represented in a transparent and flexible human readable form.

Main drawback of fuzzy decision making systems with IF-THEN rules is difficulties related with learning of parameters of decision system. Neural Networks with a certain learning algorithm can be powerful tool for learning of fuzzy systems.

So there is need to provide fuzzy decision system by learning mechanisms that can learn and adapt the fuzzy systems parameters to the changing environments and system conditions.

In Suarez and Castanon-Puga (2010) and Suarez et al. (2010b) authors show an approach to decision making system based on type-2 fuzzy inference system. In Marquez et al. (2011) a methodology is proposed based on neuro-fuzzy technique to configure a type-2 fuzzy inference system into an agent. In Mendel and Wu (2010) authors consider application of type-2 fuzzy logic to model a subjective decision-making system or perception. As it is mentioned above T2FLSs result better performance than Type-1 Fuzzy Logic Systems (T1FLSs) on the application of function approximation, modeling and decision. Based on the advantages of T2FLSs and neural networks in Castillo and Melin (2004), Lee and Lin (2005), Mendel (2001), Pan et al. (2007), Wang et al. (2004) the Type-2 neural fuzzy systems are presented to handle the system uncertainty and reduce the rule number and computation. In Pan et al. (2007) and Lee and Pan (2007) authors proposed that

© Springer International Publishing Switzerland 2014
R.A. Aliev, B.G. Guirimov, *Type-2 Fuzzy Neural Networks and Their Applications*,
DOI 10.1007/978-3-319-09072-6_5

Fig. 5.1 T2FNN system
with M rules

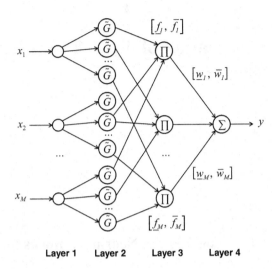

Layer 1   Layer 2   Layer 3   Layer 4

a T2FNN with asymmetric fuzzy membership functions can improve the system performance and obtain better approach ability.

In considered above works on T2FNN the structure of network was a static model. In Lee and Hu (2008) authors proposed a combining interval Type-2 fuzzy asymmetric membership functions with recurrent neural network system, called RiT2FNNA. The proposed T2FNN in Lee and Hu (2008) and Lee and Teng (2000) provides memory elements to capture system dynamic information.

Let's consider interval Type-2 fuzzy neural network decision making system.

In general, given a system input data set x_i, $i = 1, 2, \ldots, n$, and the desired output y_p, $p = 1, 2, \ldots, m$, the j-th Type-2 fuzzy rule in decision system knowledge base has the form:

$$
\begin{aligned}
R^j \quad & IF \; x_1 \;\; is \;\; \tilde{G}_1^j \, and \ldots x_n \;\; is \;\; \tilde{G}_n^j \\
& THEN \; y_1 \;\; is \;\; \tilde{w}_1^j \, and \ldots y_m \;\; is \;\; \tilde{w}_1^j, j = 1, \ldots, m
\end{aligned}
\tag{5.1}
$$

where $\tilde{G}_i j$ represents the linguistic term of the antecedent part, $\tilde{w}_p j$ represents the term of the consequent part; n and m are the numbers of the input and output dimensions, respectively. Based on the IT2FLSs, the construction of multi-input single-output (MISO) type of the T2FNN system is shown in Fig. 5.1 (Lee and Hu 2008).

Obviously, it is a static model and the structure uses interval type-2 fuzzy sets ($\tilde{G}$ and $\tilde{w}$). For simplicity we will use a commonly used two-dimensional interval type-2 Gaussian MF (Fig. 1.8b) with an interval mean in $[m_1, m_2]$ and fixed variance σ. It can be found that the IT2FNN uses the interval type-2 fuzzy sets and it implements the FLS in a four layer neural network structure.

Layer 1 nodes are input linguistic variables and layer 4 nodes are output nodes. Layer 2 nodes act as type-2 membership functions. Layer 3 nodes formulate a fuzzy rule base and the links between layers 3 and 4 function as a connectionist inference engine.

Let's describe decision processes in the type-2 neuro-fuzzy decision making system. i-th node of layer 1 is represented as:

$$O_i^{(1)} = x_i^{(1)} \tag{5.2}$$

where $x_i^{(1)}$ represents the input to the ith node of layer 1.

In layer 2, each node performs a Type-2 membership function. For Gaussian membership functions with uncertain mean the output $O_{ij}^{(2)}$ is represented as

$$O_{ij}^{(2)} = \exp\left[-\frac{1}{2} \frac{\left(O_i^{(1)} - m_{ij}\right)^2}{\left(\sigma_{ij}\right)^2} \right] = \begin{cases} \overline{O}_{ij}^{(2)} & as \quad m_{ij} = \overline{m}_{ij} \\ \underline{O}_{ij}^{(2)} & as \quad m_{ij} = \underline{m}_{ij} \end{cases} \tag{5.3}$$

where m_{ij} and σ_{ij} represent the mean and the variance respectively. The subscript "ij" indicates the jth term of the ith input $O_i^{(1)}$. The output is represented as $\left[\underline{O}_{ij}^{(2)} \overline{O}_{ij}^{(2)} \right]$.

In layer 3, the operation is the product operation:

$$O_j^{(3)} = \prod_{i=1}^n \left(O_{ij}^{(2)} \right) = \begin{cases} \overline{O}_j^{(3)} = \prod_{i=1}^n \left(\overline{w}_{ij}^{(3)} \overline{O}_{ij}^{(2)} \right) \\ \underline{O}_{ij}^{(3)} = \prod_{i=1}^n \left(\underline{w}_{ij}^{(3)} \underline{O}_{ij}^{(2)} \right) \end{cases} . \tag{5.4}$$

Links in layer 4 are used to implement the consequence matching, type-reduction and defuzzification (Lee and Lin 2005; Mendel 2001; Lee and Hu 2008):

$$y = O^{(4)} = \frac{O_R^{(4)} + O_L^{(4)}}{2}$$

where:

$$O_L^{(4)} = \sum_{j=1}^M f_j^L \underline{w}_j^{(4)} = \sum_{j=1}^L \overline{O}_j^{(3)} \underline{w}_j^{(4)} + \sum_{j=L+1}^M \underline{O}_j^{(3)} \underline{w}_j^{(4)},$$

$$O_R^{(4)} = \sum_{j=1}^M f_j^R \overline{w}_j^{(4)} = \sum_{j=1}^R \underline{O}_j^{(3)} \overline{w}_j^{(4)} + \sum_{j=R+1}^M \overline{O}_j^{(3)} \overline{w}_j^{(4)}.$$

and

$$L = \operatorname*{arg\,min}_{j \in [1, \ldots, M-1]} \left(\underline{O}_i^{(4)} \right), \qquad R = \operatorname*{arg\,max}_{j \in [1, \ldots, M-1]} \left(\overline{O}_R^{(4)} \right).$$

To calculate $Q_L^{(4)}$ and $Q_R^{(4)}$, one need to find coefficients R and L in accordance with iterative Karnik-Mendel procedures presented in Mendel (2001) and Liang

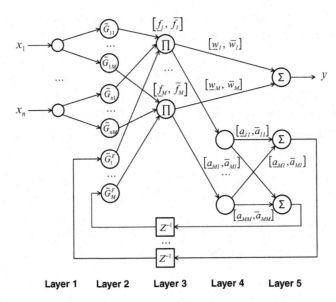

Layer 1   Layer 2   Layer 3   Layer 4   Layer 5

Fig. 5.2 Structure of the recurrent type-2 neuro-fuzzy system

and Mendel (2000). Thus, the output decision of type-2 neuro-fuzzy decision making system is determined as follows (Lee and Hu 2008):

$$y = \frac{1}{2}\left[\sum_{j=1}^{L} \overline{O}_j^{(3)} \underline{w}_j^{(4)} + \sum_{j=L+1}^{M} \underline{O}_j^{(3)} \underline{w}_j^{(4)} + \sum_{j=1}^{R} \underline{O}_j^{(3)} \overline{w}_j^{(4)} + \sum_{j=R+1}^{M} \overline{O}_j^{(3)} \overline{w}_j^{(4)}\right].$$

The structure of type-2 decision making system considered till this point (Fig. 5.1) was a static model. In the case of multi-stage dynamic decision making one needs to combine type-2 fuzzy system with recurrent neural networks system (Lee and Hu 2008). In Lee and Hu (2008) a recurrent interval Type-2 fuzzy neural network system, which provides memory element to capture system dynamic information, is suggested. The structure of the suggested in Lee and Hu (2008) system is shown in Fig. 5.2.

Thus structure is similar to the structure of type-2 neuro-fuzzy system shown in Fig. 5.1, but has an additional layer 5 (Fig. 5.2), the recurrent part of the type-2 fuzzy system.

A jth rule-base of the recurrent interval type-2 neuro-fuzzy decision making system can be described as:

$$R^j : \; IF \;\; x_1 \;\; is \;\; \widetilde{G}_{1j} \;\; and \;\; \dots \;\; x_n \;\; is \;\; \widetilde{G}_{nj} \;\; and \;\; g_j \;\; is \;\; \widetilde{C}_j^F$$

$$THEN \;\; y_1 \;\; is \;\; \widetilde{w}_1^j \;\; and \;\; \dots \;\; y_m \;\; is \;\; \widetilde{w}_m^j, \;\; g_1 \;\; is \;\; \widetilde{a}_1^j, g_2 \;\; is \;\; \widetilde{a}_2^j, \dots, \;\; and \;\; g_M \;\; is \;\; \widetilde{a}_M^j,$$

where $\widetilde{G}$ represents the linguistic term of the antecedent part, $\widetilde{w}$ and $\widetilde{a}$ represents the terms of the consequent part; and M is the total rule number.

The information propagation and the operation functions of the nodes in each layer is described in Lee and Hu (2008) in details, and, in general, is similar to procedures described for structure shown in Fig. 5.1. The additional feedback layer contains the context nodes, which is used to produce the interval variable $O_j^{(5)}$. The type-reduction and defuzzification operations as layer 4 are performed here as follows:

$$g_j(k+1) = O_j^{(5)}(k+1) = \frac{1}{2}\left[\underline{O}_j^{(5)}(k+1) + \overline{O}_j^{(5)}(k+1)\right]$$

$$\underline{O}_j^{(5)}(k+1) = \underline{a}_j^T f_L = \sum_{h=1}^{L'}\left(\overline{O}_k^{(3)}\underline{a}_{jh}\right) + \sum_{h=L'+1}^{M}\left(\underline{O}_k^{(3)}\underline{a}_{jh}\right)$$

$$\overline{O}_j^{(5)}(k+1) = \overline{a}_j^T f_R = \sum_{h=1}^{R'}\left(\underline{O}_k^{(3)}\overline{a}_{jh}\right) + \sum_{h=R'+1}^{M}\left(\overline{O}_k^{(3)}\overline{a}_{jh}\right)$$

$$L_j^F = \underset{j\in 1,...,M}{\arg\ \min}\left(\underline{O}_j^{(5)}(k+1)\right)$$

$$R_j^F = \underset{j\in 1,...,M}{\arg\ \max}\left(\overline{O}_j^{(5)}(k+1)\right)$$

It is worth to mention that each fuzzy rule has the corresponding internal variable which is used to decide the influence degree of temporal history to the current rule.

5.2   Type-2 Fuzzy Neural Networks for Forecasting

The problem that we consider in this section is forecasting petrol production for optimal scheduling of oil refinery plant (Aliev et al. 2011). In the fuzzy forecasting model we used three inputs and one output where the predicted variable is associated with the historical data $x(k)$, $x(k-1)$, and $x(k-2)$:

$$x(k+1) = F(x(k-2), x(k-1), x(k)).$$

For this example, we considered the actual daily data coming from existing petrol production unit and covering 1 month period. The fuzzy prediction model was built on the basis of type-2 fuzzy IF-THEN rules with three type-2 fuzzy variables in antecedent and one in consequent part. For the implementation of the system, the type-2 fuzzy neural network suggested in Sect. 3.6 (Fig. 3.29) with the clustering algorithm described in Sect. 4.3 and Fig. 4.6 of this book were used.

Approximately 80 % of the experimental data (selected randomly) were used for training the network and the remaining data were used for testing purposes.

Table 5.1 Cluster centers (prototypes) obtained for petrol production data

Variables\cluster centers	1	2	3	4
$x(k-2)$	[12.30,28.93]	[30.72,38.81]	[44.03,52.25]	[37.49,47.48]
$x(k-1)$	[38.02,38.59]	[40.06,41.36]	[35.71,36.24]	[37.63,38.21]
$x(k)$	[36.88,37.87]	[38.93,40.11]	[40.96,41.18]	[20.45,23.24]
$x(k+1)$	[40.48,43.61]	[15.06,19.97]	[30.67,33.40]	[46.77,49.82]

By applying the fuzzy clustering (the one considered in Sect. 4.3 of this book was used) for the available training data, it was concluded, on the basis of the selected validation criterion (4.15), that the optimal number of clusters is 4. Therefore four rules with four membership functions for each input and output variables' term-sets were used as the model's initial knowledge base.

The clustering algorithm (fuzzy clustering algorithms are described in some detail in Chap. 4 of this book) produced the cluster centers shown in Table 5.1.

The rules come in the form:

IF $x(k-2)$ is $A1$ and $x(k-1)$ is $B1$ and $x(k)$ is $C1$ THEN $x(k+1)$ is $D1$;
IF $x(k-2)$ is $A2$ and $x(k-1)$ is $B2$ and $x(k)$ is $C2$ THEN $x(k+1)$ is $D2$;
IF $x(k-2)$ is $A3$ and $x(k-1)$ is $B3$ and $x(k)$ is $C3$ THEN $x(k+1)$ is $D3$;
IF $x(k-2)$ is $A4$ and $x(k-1)$ is $B4$ and $x(k)$ is $C4$ THEN $x(k+1)$ is $D4$.

The initial type-2 fuzzy terms $A1, A2, A3, A4$ were formed from the component x $(k-2)$ of the cluster vectors 1, 2, 3, and 4, respectively:

$$A1 = [[12.30, \ 12.30], [12.30, \ 28.93], [28.93, \ 28.93]];$$
$$A2 = [[30.72, \ 30.72], [30.72, \ 38.81], [38.81, \ 38.81]];$$
$$A3 = [[44.03, \ 44.03], [44.03, \ 52.25], [52.25, \ 52.25]];$$
$$A4 = [[47.48, \ 47.48], [37.49, \ 47.48], [47.48, \ 47.48]].$$

Similarly, we formed the terms $B1, B2, B3, B4, C1, C2, C3, C4, D1, D2, D3,$ and $D4$.

The MSE values obtained for the training data and for testing data were equal to $1.2 \cdot 10^{-3}$ and 1.57, respectively.

The type-2 fuzzy terms obtained after training come as follows:

$$A1 = [[0.00, \ 7.97], [12.300, \ 25.53], [52.00, \ 59.00]];$$
$$A2 = [[4.14, \ 16.24], [20.86, \ 24.62], [47.11, \ 57.19]];$$
$$A3 = [[2.42, \ 5.63], [10.47, \ 31.87], [54.47, \ 57.60]];$$
$$A4 = [[9.23, \ 15.75], [18.79, \ 62.92], [66.56, \ 69.03]];$$

$$B1 = [[0.00, \ 1.20], [2.16, \ 24.64], [44.00, \ 52.00]];$$
$$B2 = [[42.44, \ 49.21], [57.37, \ 70.72], [75.77, \ 79.99]];$$
$$B3 = [[15.58, \ 18.32], [24.68, \ 28.75], [29.07, \ 34.99]];$$
$$B4 = [[31.42, \ 32.00], [35.07, \ 39.84], [40.03, \ 48.99]];$$

$$C1 = [[0.00, \ 23.99], [32.52, \ 40.87], [75.45, \ 77.99]];$$
$$C2 = [[7.99, \ 36.45], [57.75, \ 60.80], [64.48, \ 79.99]];$$
$$C3 = [[17.95, \ 33.72], [35.13, \ 41.98], [50.74, \ 79.99]];$$
$$C4 = [[23.08, \ 34.00], [40.36, \ 49.68], [58.17, \ 79.99]];$$

$$D1 = [[12.13, \ 12.13], [40.28, \ 75.31], [79.99, \ 79.99]];$$
$$D2 = [[12.10, \ 12.10], [34.43, \ 37.44], [41.99, \ 41.99]];$$
$$D3 = [[9.29, \ 9.29], [18.54, \ 50.42], [56.68, \ 56.68]];$$
$$D4 = [[0.00, \ 0.00], [0.00, \ 0.00], [31.01, \ 31.01]].$$

The used type-2 fuzzy clustering method (Fig. 4.6) (Aliev et al. 2011) provides better location of the cluster centers, and subsequently results in a better fuzzy rule model. This in turn allows capturing more uncertainty and deliver higher robustness against the imprecision of the data.

5.3   Type-2 Fuzzy Neural Networks for Customer Credit Evaluation

The credit risk evaluation is a major problem present in the area in banking and finance. It is extremely difficult to determine whether or not a granted loan will return in due time or will be returned at all. Many banks have developed methods to evaluate credibility of their clients before granting loans but they never have been fully satisfied with the end result. There are many studies on client evaluation using intelligent techniques. These techniques mostly involve rule based construction in which fuzzy reasoning is the leading method. The rule base is related with uncertainty implied by a variety of survey procedures being used to mine knowledge from experts. So far, there has been no research done by using type-2 fuzzy logic approach to loan assessment. Here, we look at this modeling alternative. The proposed system has been applied to a sample of real data received from a local bank.

In the considered case, the main factors taken into consideration in credit analysis are: credit history, age, net income (salary), loan amount, loan maturity, guarantors and availability of collateral.

The decision-making is based on the information about the applicant addressing the following points: net income, age, last employment period, credit history, purpose of loan, requested loan amount, loan maturity, number of guarantors, and collateral. An excerpt from the data base including the input data items together with the corresponding expert decisions is presented in Table 5.2.

To develop an intelligent system to automatically evaluate loan applicants in terms of their creditability, the experts were interviewed and the initial rule-base was formed. To make the functionality of the decision-making process transparent and trustworthy, the inference system was constructed through a collection of fuzzy IF-THEN rules, which would keep it compact and human interpretable.

Table 5.2 An excerpt from loan applicants' database

#	Net income (USD)	Age (yrs)	Last empl. period (yrs)	Credit history	Purpose of loan	Requested loan amount	Loan maturity (yrs)	Number of guarantors	Collateral	Loan request
1	1,073	29	3	Negative	Flat refurb.	3,000	36	1	N/A	Denied
2	893	32	4	Negative	Flat refurb.	3,000	36	2	N/A	Denied
3	664	25	2	Positive	Car purch.	6,000	36	2	Car	Accepted
4	1,348	34	2	Positive	Car purch.	8,000	36	2	Car	Accepted
5	250	20	0.5	Positive	Car purch.	2,000	24	2	Car	Denied
6	400	24	3	Positive	Flat refurb.	2,500	12	1	N/A	Accepted
7	140	25	1	Positive	Car purch.	1,500	30	2	Car	Denied
8	524	39	5	Positive	Flat purch.	5,000	36	2	Flat	Accepted
9	662	32	4	Positive	Flat purch.	6,500	36	1	Flat	Accepted
10	1,695	37	7	Positive	Flat purch.	15,000	24	1	Flat	Accepted

The historical database collected during a certain period of activity of the bank (with information about 460 applicants) was used to validate and optimize the initial rule-base using the proposed clustering technique. In the rules, the entries (inputs) "Net income" $(X1)$, "Age" $(X2)$, "Last employment period" $(X3)$, and "Requested loan amount" $(X6)$ were treated as type-2 fuzzy input variables. The remaining entries were treated as type-1 fuzzy or numerical variables.

The derived rule-base consisted of fuzzy rules such as those presented below:

```
IF
"Net income" is "High (X1)" and
"Age" is "Average (X2)" and
"Last employment period" is "High (X3)" and
"Credit history" is "High (X4)" and
"Purpose of loan" is "Flat purchase (X5)" and
"Requested loan amount" is "Average (X6)" and
"Maturity of loan" is "Average (X7)" and
"Number of guarantors" is "At least one (X8)" and
"Collateral" is "Don't care (X9)"

THEN "Accept" is "Very high (Y)"
```

The rule-base was aligned with the clustering results and submitted to the experts for further revision. Finally, the approved rule-base contained 34 fuzzy IF-THEN rules. The variable term-sets were defined by initial membership functions which were afterwards adjusted by the training procedure realized in the suggested Type-2 Fuzzy Neural Network (Sects. 3.6.2 and 4.3).

The performance of the obtained system was compared with the performance produced by the "standard" feed-forward neural network with nine input $(X1,\ldots, X9)$, ten neurons in the hidden layer (which was based the trial-and-error approach: the network with ten neurons in hidden layer converged better than ones with seven to nine or more than ten neurons), and one output neurons trained by means of the back-propagation learning scheme on available 460 data entries with an initial learning rate set to 0.6. Although both systems showed similar numeric results (performance), the advantage of the fuzzy system comes with respect to much higher value of trustworthiness of achieved decision due to the transparency of the underlying decision-making mechanism (Aliev et al. 2011).

5.4   Type-2 Fuzzy Neural Networks for Identification

This benchmark problem for nonlinear system identification was taken from Chen and Linkens (2004). The dynamic system is governed by the following equation:

$$x(k) = g(x(k-1), x(k-2)) + u(k) \tag{5.5}$$

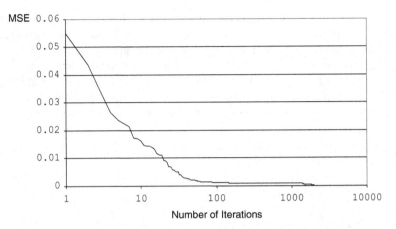

Fig. 5.3 Minimized performance index in successive iterations of the DE optimization

where:

$$g(x(k-1), x(k-2)) = \frac{x(k-1)x(k-2)(x(k-1)-0.5)}{1+x^2(k-1)+x^2(k-2)} \qquad (5.6)$$

The goal is to construct the T2FINN (refer to Sects. 3.6.2 and 4.3) for the non-linear system Eqs. 5.5 and 5.6. 200 input-output data pairs were used for training and 200 others for testing.

After applying the DE-based clustering (Sect. 4.3, Fig. 4.6), we arrive at the following five rules:

IF $x(k-2)$ is $A1$ and $x(k-1)$ is $B1$ THEN $x(k)$ is $C1$
IF $x(k-2)$ is $A2$ and $x(k-1)$ is $B2$ THEN $x(k)$ is $C2$
IF $x(k-2)$ is $A3$ and $x(k-1)$ is $B3$ THEN $x(k)$ is $C3$
IF $x(k-2)$ is $A4$ and $x(k-1)$ is $B4$ THEN $x(k)$ is $C4$
IF $x(k-2)$ is $A5$ and $x(k-1)$ is $B5$ THEN $x(k)$ is $C5$.

The progression of the DE optimization is quantified in terms of the values of the MSE reported in successive iterations, see Fig. 5.3. It becomes noticeable that the reduction of the error is quite fast and the optimization proceeds smoothly without any significant oscillations.

The type-2 fuzzy terms obtained after the training are the following:

Type-2 fuzzy terms for $x(k$-2$)$:

$$A1 = [[-29.46, -14.39], [-3.38, -3.38], [4.50, 40.91]];$$
$$A2 = [[-13.40, -9.95], [-9.89, 9.39], [9.39, 16.29]];$$
$$A3 = [[-6.12, -0.58], [1.84, 3.04], [8.37, 10.12]];$$
$$A4 = [[-12.70, -1.69], [2.93, 3.50], [6.60, 7.09]];$$
$$A5 = [[-17.17, -15.70], [-2.12, -2.03], [-2.03, 5.40]]$$

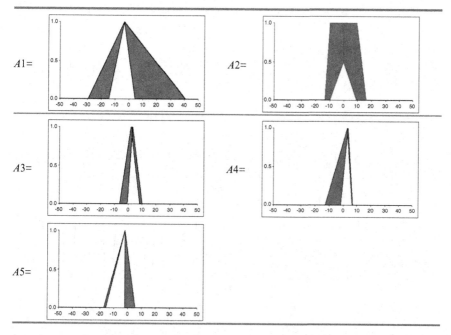

Fig. 5.4 Final FOUs

Type-2 fuzzy terms for $x(k$-$1)$:

$$B1 = [[-20.98, -20.72], [3.18, 4.91], [35.59, 38.01]];$$
$$B2 = [[-16.90, -1.38], [-0.83, 2.19], [2.43, 6.73]];$$
$$B3 = [[-17.95, -8.29], [-2.78, -1.58], [0.10, 1.59]];$$
$$B4 = [[-2.03, -1.37], [-0.95, 1.40], [1.41, 1.52]];$$
$$B5 = [[-8.96, -3.84], [-3.57, -0.30], [10.84, 15.49]]$$

Type-2 fuzzy terms for $x(k)$:

$$C1 = [[-6.77, -6.77], [-6.74, -1.86], [15.46, 15.46]];$$
$$C2 = [[-5.92, -5.92], [-2.01, 9.87], [13.53, 13.53]];$$
$$C3 = [[-10.28, -10.28], [-0.81, 2.08], [22.32, 22.32]];$$
$$C4 = [[-15.85, -15.85], [-4.91, 2.10], [9.80, 9.80]];$$
$$C5 = [[-19.73, -19.73], [-14.11, -1.69], [8.39, 8.39]]$$

For further visualization, the obtained FOUs for the type-2 fuzzy terms $A1$, $A2$, $A3$, $A4$, and $A5$ after training are illustrated in Fig. 5.4.

The values of the MSE obtained on the training data and on the testing data were equal to $2.11 \cdot 10^{-4}$ and $2.84 \cdot 10^{-4}$, respectively. For comparison, as reported in Chen and Linkens (2004), the MSE for the same problem was $1.90 \cdot 10^{-4}$ (training data)

Table 5.3 MSE values produced by different versions of the network

	MSE (train data)	MSE (test data)
Fuzzy model (Chen and Linkens 2004) (type-1)	$1.90 \cdot 10^{-4}$	$3.80 \cdot 10^{-4}$
Type-1 FINN	$5.40 \cdot 10^{-4}$	$9.50 \cdot 10^{-4}$
Type-2 FINN	$2.11 \cdot 10^{-4}$	$2.84 \cdot 10^{-4}$

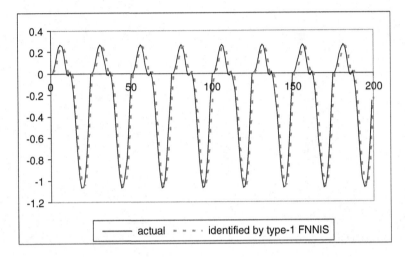

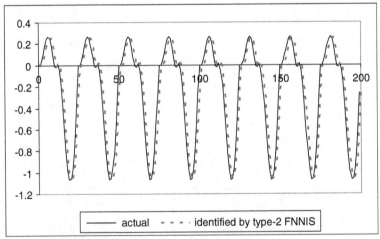

Fig. 5.5 The actual system response and the output of the model

and $3.80 \cdot 10^{-4}$ (testing data). For the type-1 fuzzy based system, the best result obtained for the same problem was $9.50 \cdot 10^{-4}$ on the testing data. All the results are concisely summarized in Table 5.3.

The comparison of the actual response of the system and the output of the model is included in Fig. 5.5 (the results are shown for testing data).

5.5   Type-2 Fuzzy Neural Networks for Control

A typical structure of a type-2 fuzzy neural network based control system is shown in Fig. 5.6. In the figure $y(k)$ represents the output signal of the plant, $g(k)$ is the set-point signal, $e(k)$ and $\Delta e(k)$ are the error and its change, respectively. Unit D outputs the error difference.

In this particular example the fuzzy wavelet neural network described in Sect. 3.8 (Abiyev et al. 2013) is used for the control of a linear time varying plant (Zhang et al. 1998).

$$
\begin{aligned}
(s^2 + sa_1(t) + a_2(t))y_p &= u_p, \\
a_1(t) &= \frac{0.1t}{t+1}, \\
a_2(t) &= \begin{cases}
0.3, & t \in [0, 40); \\
0.1, & t \in [40, 60), \\
0.6, & t \in [60, 85), \\
0.3, & t \in [85, +\infty).
\end{cases}
\end{aligned}
\tag{5.7}
$$

The type-2 FWNN (described in some detail in Sect. 3.8) with three fuzzy IF-THEN rules was considered for control of the plant (Eq. 5.7). The network has in total 33 parameters to adjust during the training. The initial values of the parameters representing the Gaussian membership functions were set as random numbers spread in the interval $[-5, 15]$ and the parameters ω were generated from $[-1, 1]$.

The gradient based training algorithm for this type of FWNN developed in Abiyev et al. (2013) was used to update the initial parameters for different plant reference functions. The training converged after approximately after 100 iterations. Figure 5.7 illustrates the progress of training of the considered FWNN. The RMSE reached was 0.6846.

Figure 5.8 demonstrates the time response characteristics of the network after the training. The type-2 the network successfully controlled the plant with abrupt changing characteristics $a_2(t)$, exceeding in performance other approaches (Zhang et al. 1998).

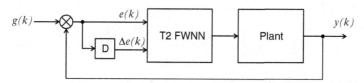

Fig. 5.6 A structure of a type-2 fuzzy neural network based control system

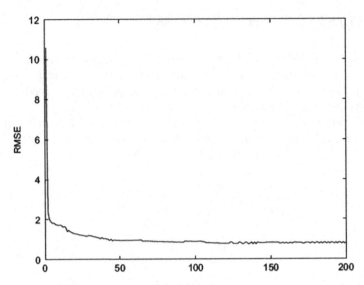

Fig. 5.7 Training performance of the considered FWNN

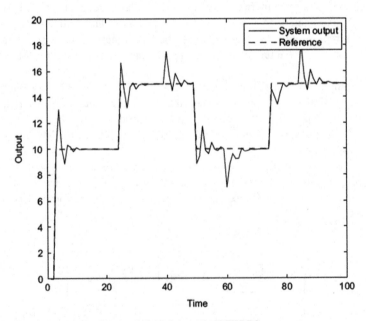

Fig. 5.8 The time response characteristics of the trained FWNN

5.6   Type-2 Fuzzy Neural Networks and Computing with Words (CWW)

As we move further into the age of intelligent machine revolutionary natural language (NL) modeling, NL computation will grow in visibility and importance. Foundation of computation with information described in natural language is paradigm of computing with words (CWW) suggested by Prof. L. Zadeh (2001, 2002). As Zadeh mentions "CWW is a methodology in which the objects of computation are words and propositions drawn from a natural language".

In Mendel (2007) the author explains Zadeh's CWW paradigm in details and argues that IT2 FS models for words are more useful to capture uncertainties about a word.

In Zadeh (1996) Zadeh first equates fuzzy logic (FL) with *computing with words* (CWW). It is obvious that computers do not actually compute using words. In CWW computers are activated by words, which are converted into a mathematical representation using fuzzy sets (FSs) (Zadeh 1965, 1975). Then these FSs are mapped by means of a CWW engine into some other FSs, after which the latter are converted back into words (Mendel 2007).

As it is shown in Mendel (2007), because words can mean different things to different people, it is important to use an FS model that lets us capture word uncertainties. In literature usually they use two possible choices, a Type-1 FS or an interval Type-2 FSs (Mendel 2001, 2003, 2007).

The general structure of CWW is shown in Fig. 5.9.

Natural language processing and computation based on CWW paradigm consist of three steps: precisiation, reasoning through CWW engine, converting to NL (Abdullayev and Aliyeva 2006; Aliev and Aliev 2001).

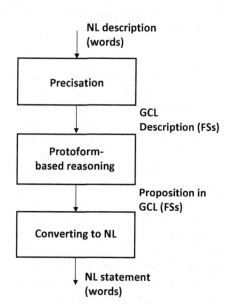

Fig. 5.9   Structure of CWW

Figure 5.9 provides schematic description of natural language processing and computation by CWW paradigm.

First module. Collection of propositions expressed in natural language and translation (explication) them into a formal computer manipulability language. In Zadeh's theory it is called Generalized Constraint Language (GCL). In this module we need to provide ways for expressing many types of natural language statements within the general constraint language.

This module transforms linguistic perceptions, i.e. words, into FSs that activate a CWW engine.

Second module. It is a goal directed manipulation of these propositions. One promising direction in this reference process is what Zadeh calls protoform-based reasoning (Zadeh 2001).

Many choices have to be made when designing the CWW engine. The CWW engine may take the form of IF-THEN rules (Mendel 2001; Klir and Yuan 1995; Aliev and Aliev 2001), linguistic weighted average (Wu and Mendel 2006), linguistic summarizations (Kacprzyk and Yager 2001; Kacprzyk and Zadrozny 2005a, b; Kacprzyk et al. 2000, 2006; Yager 1982), etc., for which the established mathematics of fuzzy sets provides the transformation from the input FSs to the output FSs.

Below we will briefly characterize each of these modules (Aliev et al. 2004; Zadeh 1981; Wang 2001). Zadeh's approach to the representation of fuzzy constraints is based on test-score semantics. According to this semantics, a proposition of natural language is considered as a network of fuzzy constraints. The latter can be represented as X is R, where R is a constraining fuzzy relation and X is the constrained variable. As it is shown in Zadeh (2000), the expression in question is the canonical form of p. The canonical form is necessary for explicitation of the fuzzy constraints in evidence which is implicit in proposition. It is necessary to note that a constraint expressed as a canonical form may be conditional:

$$\text{IF } X \text{ is } R \text{ THEN is } S. \qquad (5.8)$$

The constraints in question are the basic constraints. For proposition meaning representation, in general case, the generalized constraints are used, which contain the basic constraints as a special case. A generalized constraint is represented as X isr R (Zadeh 2000). Here r is a discrete variable, which defines the way in which R constraints X. Depending on the type of value of r in the practice of CW, the following constraints are possible (Zadeh 2000): equality, $= e$; possibilistic, $r = blank$; probabilistic etc. A very common choice, which we follow, is to use possibilistic constraints.

Words in the CWW paradigm can be modeled by Type-1 (T1) fuzzy sets or their extension, Type-2 (T2) FSs. CWW using T1 FSs has been studied by many researchers (Lawry 2001; Lawry et al. 2003), however, as claimed in (Mendel and Wu 2007), "Words mean different things to different people, and so are uncertain. We therefore need in FS a model for a word that has the potential to capture its uncertainties, and an interval T2 FS (IT2 FS) should be used as a model of a word". Consequently, IT2 FSs can be used to model words. In second module by use of constraint propagation, the antecedent constraints are transformed into consequent constraints.

It is necessary to note that fuzzy constraint propagation rules in principle are the same as fuzzy inference rules. With this in mind, the well-known compositional rule, generalized Modus Ponens, syllogistic rule, and others are commonly used.

Usually, a CWW engine takes the form of IF-THEN rules. When the CWW Engine is a set of IF-THEN rules, one can optimize the parameters of antecedent and consequent membership functions, to determine the number of significant rules. As it was shown in Sect. 3.6.2 to this end one can use interval Type-2 FNN. Outputs of IT2FNN then will be interval Type-2 fuzzy sets.

Third module. Provides converting a statement in Generalized language (GCL) into an appropriate statement in natural language (NL) using linguistic approximation procedure.

Converting module is based on retranslation process which performs linguistic approximation (Yager 2004). In fact, the main task of this third module is due to which we obtain the proposition in a natural language statement.

Let us have a collection of terms from a natural language i.e. natural language vocabulary. Formally, the problem of converting into NL is replacing of proposition V is A by V is L, where L is some element from the natural language vocabulary. Each fuzzy subset corresponds to a unique natural language term from the vocabulary. The process of retranslation is then substituting the proposition V is F by V is A, where F is some element from fuzzy sets collection and then expressing the output as V is L, where L is the linguistic term associated with F. The key issue in this process is the substitution of V is F for V is A (Yager 2004).

A degree to which the statement V is A entails the statement V is F (D) can be determined as (Yager 2004):

$$D(A \subseteq F) = \underset{x \in X}{Min} \ [I(A(x), F(x))] \tag{5.9}$$

where the operator I is an implication operator.

The closeness of F to A should be related to the distance between these two sets. The closeness of A and F may be measured by the difference

$$\Delta_j = |A(x_j) - F(x_j)|.$$

In particular, Hamming distance is determined as:

$$D_1(A, F) = \sum\nolimits_{j=1}^{n} \Delta_j \tag{5.10}$$

References

Abdullayev KM, Aliyeva AS (2006) Applications of fuzzy logic to natural language modeling. 7th International Conference on Application of Fuzzy Systems and Soft Computing (ICAFS-2006). Siegen, Germany.

Abiyev R, Kaynak O, Kayacan E (2013) A type-2 fuzzy wavelet neural network for system identification and control. Journal of the Franklin Institute 350. 1658–1685.

Aliev RA, Aliev RR (2001) Soft computing and its application. World Scientific, New Jersey, London, Singapore, Hong Kong.

Aliev RA, Fazlollahi B, Aliev RR (2004) Soft computing and its applications in business and economics. Springer-Verlag Berlin Heidelberg.

Aliev RA, Pedrycz W, Guirimov B, Aliev RR, Ilhan U, Babagil M, Mammadli S (2011) Type-2 fuzzy neural networks with fuzzy clustering and differential evolution optimization. Information Sciences, Volume 181 Issue 9, 1591–1608.

Castillo O, Melin P (2004) Adaptive noise cancellation using type-2 fuzzy logic and neural networks. IEEE International Conf. on Fuzzy Systems, Vol. 2, 1093–1098, 2004.

Chen MY, Linkens DA (2004) Rule-base self-generation and simplification for data-driven fuzzy models J. Fuzzy Sets and Systems 142, 243–265.

Kacprzyk J, Yager RR (2001) Linguistic summaries of data using fuzzy logic. Int. J. of General Systems, vol. 30, 33–154.

Kacprzyk J, Zadrozny S (2005a) Linguistic database summaries and their protoforms: toward natural language based knowledge discovery tools. Information Sciences, vol. 173, 281–304.

Kacprzyk J, Zadrozny S (2005b) Fuzzy linguistic data summaries as a human consistent, user adaptable solution to data mining. In: Gabrys B, Leiviska K, Strackeljan J (eds.) Do Smart Adaptive Systems Exist? 321–339, Springer, Berlin Heidelberg New York.

Kacprzyk J, Yager RR, Zadrozny S (2000) A fuzzy logic based approach to linguistic summaries of databases. Int'l. J. of Applied Mathematics and Computer Science, vol. 10, 813–834.

Kacprzyk J, Wilbik A, Zadrozny S (2006) Linguistic summaries of time series via a quantifier based aggregation using the Sugeno integral. In: Proc. FUZZ-IEEE 2006, 3610–3616, Vancouver, BC, Canada, July 2006.

Klir GJ, Yuan B (1995) Fuzzy sets and fuzzy logic: Theory and applications. Prentice-Hall, Upper Saddle River, NJ.

Lawry J (2001) An alternative to computing with words. Int. J. of Uncertainty, Fuzziness and Knowledge-Based Systems, vol. 9, Suppl., 3–16.

Lawry J, Shanahan J, Ralescu A (eds) (2003) Modeling with words. Lecture Notes in Artificial Intelligence, 2873, Springer, New York.

Lee CH, Hu TW (2008) Recurrent interval type-2 fuzzy neural network using asymmetric membership functions. In: Xiaolin Hu and P. Balasubramaniam (eds) Recurrent Neural Networks. 978-953-7619-08-4, InTech.

Lee CH, Lin YC (2005) An adaptive type-2 fuzzy neural controller for nonlinear uncertain systems. International Journal of Control and Intelligent, Vol. 12, No. 1, 41–50, 2005.

Lee CH, Pan HY (2007) Enhancing the performance of neural fuzzy systems using asymmetric membership functions. Revised in Fuzzy Sets and Systems, 2007.

Lee CH, Teng CC (2000) Identification and Control of Dynamic Systems Using Recurrent Fuzzy Neural Networks, IEEE Trans. on Fuzzy Systems, Vol. 8, No. 4, 349–366, 2000.

Liang AQ, Mendel JM (2000) Interval type-2 fuzzy logic systems: theory and design. IEEE Trans. on Fuzzy Systems, Vol. 8, No. 5, 535–550, 2000.

Marquez BY, Castanon-Puga M, Castro JR, and Suarez ED (2011) A distributed agency methodology applied to complex social system, a multi-dimensional approach. In Zhang, R., Cordeiro, J., Li, X., Zhang, Z., and Zhang, J., editors, Proceedings of the 13th International Conference on Enterprise Information Systems, Volume 1, Beijing, China, 8–11 June, 2011., pages 204–209. SciTe Press.

Mendel JM (2001) Uncertain rule-based fuzzy logic systems: Introduction and new directions. Prentice-Hall, Upper-Saddle River, NJ.

Mendel JM (2003) Type-2 fuzzy sets: some questions and answers. IEEE Connections, Newsletter of the IEEE Neural Networks Society, vol. 1, 10–13.

Mendel JM (2007) Computing with words: Zadeh, Turing, Popper and Occam. IEEE Computational Intelligence Magazine November 2007.

Mendel JM, Wu H (2007) Type-2 fuzzistics for non-symmetric interval type-2 fuzzy sets: Forward problems. IEEE Trans. on Fuzzy Systems, vol. 15.

Mendel JM, Wu D (2010) Perceptual computing, aiding people in making subjective judgments. IEEPress.

Pan HY, Lee CH, Chang F K, Chang SK (2007) Construction of asymmetric type-2 fuzzy membership functions and application in time series prediction. In: International Conf. on Machine Learning and Cybernetics, Vol. 4, 2024–2030, 2007.

Suarez ED. and Castanon-Puga M (2010) Distributed agency, a simulation language for describing social phenomena. In: IV Edition of Epistemological Perspectives on Simulation, Hamburg, Germany. The European Social Simulation Association.

Suarez ED, Castanon-Puga M, Flores DL, Rodriguez-Diaz A, Castro JR, Gaxiola-Pacheco C, Gonzalez-Fuentes M (2010b) A multi-layered agency analysis of voting models. In: The 3rd World Congress on Social Simulation WCSS2010, Kassel, Germany.

Wang P (2001) Computing with words. New York, John Wiley& Sons, 2001, 448 p.

Wang CH, Cheng CS, Lee TT (2004) Dynamical optimal training for interval type-2 fuzzy neural network (T2FNN). IEEE Trans. on Systems, Man, Cybernetics Part-B, Vol. 34, No. 3, 1462–1477, 2004.

Wu D, Mendel JM (2006) The linguistic weighted average. In: Proc. FUZZ-IEEE 2006, 3030–3037, Vancouver, CA, July 2006.

Yager RR (1982) A new approach to the summarization of data. Information Sciences, vol. 28, 69–86.

Yager RR (2004) On the retranslation process in Zadeh's paradigm of the computing with words. IEEE Transactions on Systems, Man, and Cybernetics, vol. 34, No. 2, April 2004.

Zadeh LA (1965) Fuzzy Sets. Information and Control, vol. 8, 338–353.

Zadeh L (1975) The concept of a linguistic variable and its application to approximate reasoning–1. Information Sciences, vol. 8, 199–249.

Zadeh LA (1981) Test-score semantics for natural languages and meaning representation via PRUF. In: Rieger B (eds) Empirical Semantics. Germany: Brockmyer, 198–211.

Zadeh LA (1996) Fuzzy logic = computing with words. IEEE Trans. on Fuzzy Systems, vol. 4, 103–111.

Zadeh LA (2000) From computing with numbers to computing with words – from manipulation of measurements to manipulation of perceptions. In: Hampel R, Wagenknecht M, Cahker N (eds) Fuzzy Control Theory and Practice. Physica-Verlag. A Springer-Verlag Company, 3–38.

Zadeh, LA (2001) From computing with numbers to computing with words-from manipulation of measurements to manipulation of perception. In: Wang P (ed) Computing with Words. John Wiley & Sons. Inc., 35–67.

Zadeh LA (2002) Toward a perception-based theory of probabilistic reasoning with imprecise probabilities J. Statist. Plann. Infer., vol. 105, 233–264.

Zhang CJ, Shao C, Chai TY (1998) Indirect adaptive control for a class of linear time-varying plants, IEE Proceedings — Control Theory and Applications 145 (2) 141–149.

Index

© Springer International Publishing Switzerland 2014
R.A. Aliev, B.G. Guirimov, *Type-2 Fuzzy Neural Networks and Their Applications*,
DOI 10.1007/978-3-319-09072-6

Printed in the United States
By Bookmasters